# A BIBLIOGRAPHY OF GILBERT WHITE OF SELBORNE

GILBERT WHITE IN HIS YOUTH
*From a painting formerly at Knebworth Hall.*

# A BIBLIOGRAPHY OF GILBERT WHITE

## THE NATURALIST & ANTIQUARIAN OF SELBORNE

With a Biography and a descriptive
Account of the Village of Selborne

BY
EDWARD A. MARTIN, F.G.S.
Author of *Life in a Sussex Windmill; Dewponds: History, Observation and Experiment*, etc.

*Seal of Selborne Priory*

Reprinted
DAWSONS OF PALL MALL
FOLKESTONE AND LONDON
1970

*First published 1897*
*Revised edition 1934*
*Reprinted by permission*
*of the Grantors 1970*

Dawsons of Pall Mall
Cannon House
Folkestone, Kent, England
ISBN: 0 7129 0470 0



*Printed by Unwin Brothers Limited*
*The Gresham Press, Old Woking, Surrey, England*
*A Member of the Staples Printing Group*

# FOREWORD

A few words would seem to be necessary by way of introduction to this enlarged and revised edition of a work that was issued over thirty years ago. Since then many new facts in connection with the life of the great naturalist have come to light, and my former book has thus become out-of-date. The biographical portions have been retained; but for the most part the book is of a bibliographical nature, that portion dealing with the various editions of Gilbert White's great work forming the major part of the text. I cannot say that there are no other editions than those I have dealt with, but it is believed that the work is almost complete in this respect.

The arrangement in this edition has been altered to some extent. It was thought that it would be more convenient if successive editions by the same editor followed each other. The result is that strict chronological order is absent, since editions by various editors frequently issued from different presses concurrently. This rearrangement does not refer, however, to the many reprints of the last thirty years. Some of these have been issued with the same notes to the text that did duty fifty or more years ago, no attempt having been made to bring them up to date. It seems scarcely fair to such editors as Blythe, Jesse, Jardine, Brown, and Davies to reproduce now their notes without correction or comment.

For verification of some of my information I have referred to Holt-White's *Life and Letters of Gilbert White*, and my acknowledgements to this classical work are hereby tendered. I am also indebted to Mr. Wilfred Mark Webb, F.L.S., honorary secretary of the Selborne Society, and to Mr. H. H. Poole, for their valuable advice and assistance.

E. A. M.

# CONTENTS


| | | |
|---|---|---|
| Foreword | | v |
| List of Illustrations | | viii |
| CHAPTER | | PAGE |
| I. | The Naturalist | 1 |
| II. | His Book—"The Natural History and Antiquities of Selborne" | 24 |
| III. | The Village of Selborne | 34 |
| IV. | "The Wakes" and Its Associations. Some Relics of the Naturalist | 57 |
| V. | The Testimony of Men | 80 |
| VI. | Bibliography; First Edition, 1789 | 90 |
| VII. | Bibliography; Aikin and Markwick, Mitford, Jardine (in part), and Rennie | 98 |
| VIII. | Bibliography; Lady Dover, S.P.C.K., and Thomas Brown | 119 |
| IX. | Bibliography; Blyth, Bennett, and Jenyns | 129 |
| X. | Bibliography; Jesse and Jardine | 137 |
| XI. | Bibliography; J. G. Wood, Bell and Daldy, Harting, Bell, and Christopher Davies | 144 |
| XII. | Bibliography; Frank Buckland | 160 |
| XIII. | Bibliography; Haweis, Jefferies, Burroughs, and Grant Allen | 166 |
| XIV. | Bibliography; Pollard, Bowdler Sharpe, Weekes, Miall and Fowler, Kearton, and others | 176 |
| | Additional Bibliography | 189 |

## LIST OF ILLUSTRATIONS


Gilbert White in his youth — frontispiece
Seal of Selborne Priory — title-page
"The Wakes" from the rear as it was in White's time — p. 6
Pen-and-Ink sketches of Gilbert White — facing p. 16
Selborne Street showing "The Wakes" with its old front — p. 22
Selborne Church from The South East — facing p. 38
Garnetted Wall of Church, studded with "Tenpenny Nails" — p. 39
Gilbert White's Gravestone — p. 40
Trees planted by Gilbert White — facing p. 42
Encaustic Tiles from The Priory — p. 45
The Wishing Stone at the top of the Zigzag — facing p. 46
Cottages bordering on the Well-head Stream — p. 49
Where the Well-head Stream crosses the Road — p. 50
Thatched Cottages in Selborne Street — facing p. 50
View from The Hanger — facing p. 52
Between Nore Hill and the Hanger Hill — p. 56
"Delves House," Ringmer, near Lewes — facing p. 74
Stone Coffin from Selborne Priory — p. 79

# CHAPTER I

## THE NATURALIST

GILBERT WHITE, the author of *The Natural History and Antiquities of Selborne*, was born on July 18th, 1720, in an old-fashioned vicarage which formerly stood on the site of the present vicarage at Selborne.

"He was the eldest son of John White,* of Selborne," his nephew, John, wrote nine years after his death, "and of Anne,† daughter of Thomas Holt,‡ rector of Streatham, in Surrey."

John White, the father, a barrister-at-law, and a Justice of the Peace for Hants, was at the time staying with his wife at the vicarage, then in the occupation of Gilbert White, the vicar of Selborne and grandfather of the naturalist. After young Gilbert's birth, John and his wife removed to Compton, a pretty village lying under the shelter of the Hog's Back. Here they resided until 1727, when they removed to East Harting, where Mrs. Holt was residing, and thence back again to Selborne three or four years later.

"He received his school education at Basingstoke under the Rev. Thomas Warton, vicar of that place. He was admitted at Oriel College, Oxford, in December, 1739, and took his degree of Bachelor of Arts in June, 1743. In March, 1744, he was elected Fellow of his College. He became Master of Arts in October, 1746. Being of unambitious temper, and strongly attached to the charms of rural scenery, he early fixed his residence (1755), in his native village, where he spent the greater part of his life in literary occupations, and especially in the study of nature. This he followed with

*d. 1758. †d. 1739, age 46. ‡d. 1710.

a patient assiduity, and a mind ever open to the lessons of piety and benevolence, which such a study is so well calculated to afford. Though several occasions offered of settling upon a college living, he could never persuade himself to quit the beloved spot, which was indeed a peculiarly happy situation for an observer. He was much esteemed by a select society of intelligent and worthy friends, to whom he paid occasional visits. Thus his days passed, tranquil and serene, with scarcely any other vicissitudes than those of the seasons, till they closed at a mature age on June 26th, 1793."*

Gilbert White took deacon's orders in 1747, in his 27th year, becoming curate to his uncle Charles, at Swarraton. He this year suffered from a severe attack of small-pox, being repeatedly bled, and this may have left its marks upon him for years after. He was ordained priest in the Chapel in Spring Gardens, London, two years later, and was appointed curate of Selborne in Oct, 1751. He resigned this to enter on his proctorship at Oriel College in April of the next year. In 1753 he became curate at Durley, near Bishops Waltham, for a year and a half, but in 1755 returned to Selborne in his thirty-fifth year, and on the death of his father in 1758, resided at "The Wakes." He became the owner of the house in 1763 and here he resided until his own death in 1793. He was appointed curate at Faringdon in 1761, and retained the curacy at Faringdon until 1784, when he again became curate at Selborne. Although never vicar in his own parish, yet he did accept one living, viz., at Moreton-Pinkney, in Northamp-

*John White, in edition of 1802.

tonshire, in 1757, an appointment in which he gave neither residence nor service.

Gilbert White was the eldest of a family of eleven children, which followed the union of John White and Anne Holt. Of those who reached adult age, the most important next to Gilbert is perhaps Benjamin, the natural history publisher, through whom it is probable he was brought into correspondence with the two recipients of his Letters. Benjamin was the publisher of the original edition of *The Natural History and Antiquities of Selborne* in 1789. His premises, the "Horace's Head," in Fleet Street, have long since given place to the building now occupied by the Norwich Life and Fire Insurance Society. The second edition, of 1802, was published by John, son of Benjamin. John, a brother of Gilbert, was at one time chaplain at Gibraltar, and afterwards vicar at Blackburn. A third brother was Thomas White, who was a Fellow of the Royal Society, and it is to his influence with Gilbert that we are indebted for the gathering of the correspondence together in the form of a book. We are told that it was he who finally persuaded Gilbert to publish his part of the correspondence, in order that the results of the observations he made should not be lost to mankind. Thomas afterwards reviewed the book in the *Gentleman's Magazine*.

Another brother was Henry, who also entered the clerical profession. He became vicar of Fyfield, near Andover, from whence Gilbert, when visiting his brother, dated one of his letters.

In order fully to appreciate the period in which a man

has lived, it is convenient to fix on one's memory certain contemporaneous events of which he must have been cognisant, marking stage by stage the various resting-places in his life. White was born in the year 1720. This was in the sixth year of George the First's reign. When a young man of 25, the rebellion of the young Pretender took place, and we cannot doubt but that the excitement attendant on such an event penetrated even to the rural district of Selborne. He witnessed the change in the calendar in 1752, but as he had not then commenced to write his natural history letters, the dates as they appear on the letters are the dates in our own "new style."

Clive commenced his victories in India in 1757, one hundred years before the outbreak of the Indian Mutiny, and at this time Gilbert had reached the mature age of 37. He was 40 years old when the success of Pitt in his foreign and colonial policy was at its height, and in the same year (1760), George the Third commenced his long and eventful reign. He may have watched the gradual, though, temporary decadence of British arms, culminating in that unnatural strife between two kindred nations which resulted in England and America being torn roughly asunder. He would have heard in his 64th year, with mingled feelings of regret and shame, of the Declaration of Independence, and in 1789, when he published his book at the age of 69, he would have heard the echoes wafted afar, even to sleepy Selborne, of the fearful doings which were characterising the early excesses of the French Revolution. In this respect it has been well remarked by Professor Henry Morley, that "he was more concerned with the course of events in a martin's nest than with the crash of empires" which was going on around.

Such then were some of the principal events of which he would have heard during the space of his earthly life.

It has been surmised, and possibly with a great deal of truth, that White's taste may have been turned in the direction of natural history studies by association in his youthful days with Dr. Stephen Hales, who was, as we are informed in some letters first associated with Gilbert White's book in Harting's second edition of 1876 (Bickers), White's "most valued friend," and who was well known also to his grandfather and father. White spoke of him in this way to his friend, Robert Marsham, in 1790, twenty-nine years after Dr. Hales' death, when apparently the influences and impressions which he received from his former friend were still fresh and green.

In another letter to Marsham, he again recurs to the man whose esteem and friendship they had chanced to find was mutual, and referred to the numerous inventions and experiments with which Dr. Hales delighted to entertain himself and others. Dr. Hales was born in 1677, and died in 1761. He published a work on *Vegetable Staticks,* and another on *Ventilators,* the former being translated into French by Buffon, and into Italian, German, and Dutch. He was elected Fellow of the Royal Society in 1717, and published various papers in the *Philosophical Transactions.*

In attempting to gauge Gilbert White's career as a naturalist, it is necessary not only to take into account the actual amount of work which he accomplished, and which showed itself towards the end of his life in the form of a printed record of his observations, but also to consider that work in the light of the limited amount of knowledge in the natural history world then extant and attainable to a village recluse, such as it must be admitted White was. Our judgment of such a work must not be formed from the same standpoint as it would be, if the work were to appear in public now for the first time. On the other hand, from its depth of observation, and the diversity of subjects dealt with,

"THE WAKES" FROM THE REAR AS IT WAS IN WHITE'S TIME.

the minute details which characterise the observations, and the well-reasoned conclusions to which the author arrives, it could not fail, even now, to make a lasting impression on the natural history literature of the day. Appearing, however, as it did, in the comparatively barren natural history period of the eighteenth century, it showed itself to be the result of the workings, during a long life, of a brain in which dwelt the true pertinacity and perseverance of a man of genius; a genius, however, in which also was found a distinct talent for business-like habits, which allowed, nay, compelled, its owner to place on almost daily record, for the benefit of others, the results of his observations. His was a genius which for many years fed and thrived upon its own exercise, neither expecting nor wanting acknowledgment or gratitude from the world which surrounded it, the accomplishment of its purpose being the best and the most satisfying encouragement which it could wish to receive. Although the subjects dealt with by White may in some cases appear trivial, and although the whole material of the book may seem in the eyes of some to be of a very simple nature, yet we must not be led away to think that the author's world was altogether narrow and simple, or that his mental horizon was bounded by the limits of his native village. Recluse though he was for a great part of each year, the influences of his education at Oxford were sufficient to give his mind the opportunities of ample expansion, and the keen single-eyed love for Nature which he brought to bear on her problems was regulated by a shrewd common sense, which almost forbade him from risking his reputation in support of theory, whilst there was so much in the region of fact still to be discovered.

There is, though, one theory which was very dear to his heart in support of which he was extremely anxious to weigh all evidence as favourably as possible. It was in regard to the hybernation of swallows in this country that he approached perilously near to the shoals and quicksands of unstable theory, and yet the evidence which he now and again brought forward, and the manner in which he did so, showed that either he himself had not quite been convinced of its truth, or else his nature was still too timid to advance a theory founded on observations which better-known naturalists had not yet themselves made.

The observation, rather than the study of Nature, was White's peculiar characteristic, and this habit was with him wherever he might roam.

We have evidence from his letters that he was as observant as ever when visiting his apparently few friends at such places as Fyfield, near Andover, Ringmer, near Lewes, and South Lambeth. If, as we surmise, his friends outside of the White family were few, and that these held him in great estimation—as we are told by his nephew was the case—we may perhaps feel inclined to judge that he was not one who cultivated social intercourse, and that perhaps something of the cordiality between the society of one country parish and another, which is so important and necessary an adjunct of country life, was to a certain extent lacking in Gilbert White.

Whatever, however, he may have lacked in this respect, was amply compensated by his bearing towards the souls of the parish which had been committed to his charge. The words of his nephew Edmund White, quoted by Professor Bell, are evidence of the gentle and philanthropic manner which he bore to his poorer neighbours. After referring to his "pure character and religious principles," he adds that White possessed "the peculiarity of attaching to him all and every age, particularly young people, who listened with delight to his instructive tales." In the grave debates which took place on the Plestor amongst the elderly people, or in riveting the attention of the children who played around, we can well believe how prominent and necessary a character he was to the generation with whom he lived. His nephew also says, "If I were called upon to say in what point he particularly excelled, I think I should say it was in his mode of addressing his poorer neighbours. His kind philanthropy, and his charitable wishes towards them, and his humane enquiries, always made them feel that he was their true friend."

It is always interesting to have access to facts concerning the life of a famous man, other than as regards his life in the particular walk in which he may be distinguished. Things are sometimes discovered which although trivial in themselves, yet act as the straws showing the various currents which underlie his public character. Amongst such may be cited an anecdote attesting his popularity amongst his parishioners. He was dining one day at a farm-house on bacon and beans, when his hostess was good enough to do him the dubious kindness of strewing sugar on his plate. Upon his checking this exercise of her good nature, she replied "Can't be too good for you, Sir, can't be too good for you," and this at a time when sugar must have been a far greater luxury than we are apt to regard it now.

In gauging the character of the naturalist I do not think we must be too much carried away with the idea that with all his excellent qualities he was a man of so weak a disposition, and with so single an eye in his duties and in his hobbies, that he was incapacitated for the stern business of life. Doubtless he had an abounding stock of healthy vigorous innocence, such as all good men possess, but we must not thereby take it for granted that he was the unsophisticated man which he might have been, had he been quite ignorant of the experiences of life outside his own rural parish. His experiences in London enabled him to speak many times of Fleet Street and South Lambeth, both of which were familiar to him. His early life at Oxford ,and his constant visits afterwards to his *Alma Mater*, bid us avoid a conclusion other than that he was familiar with many sides of town life. True manliness was in as much estimation in White's days as in our own, and we may be sure he would not have received the boisterous welcome of the undergraduates at Oxford so often and apparently so gladly, were he of any type but that of a true, manly, and in a sense muscular, English gentleman.

It has very fairly been said that to a great extent the sporting instincts of our grandfathers have given place nowadays to the taste for studies in natural history. It is certain that many old sporting books are very good histories of the natural habits and the varieties of the creatures dealt with in their pages. White tells us distinctly that he was not a sportsman, but for all that there are many references which shew us that the sporting world still held a good deal of his interest, although he well knew the bad moral effects which waited on certain forms of so-called sport. (See Letter VII to Pennant.)

Great, too, as was his love for the feathered creation, he himself was led to see that wholesale protection of every species of bird was indefensible, if the market-garden were to give forth its fruits in due season; so we find certain passages in his diaries which show us that even he was compelled to take measures to destroy the superfluity of bird-life of which Selborne boasted. He was no single-minded enthusiast. He was moderate in his views and actions all round.

It has been said that little or no romance seems to have entered into the period of White's earthly pilgrimage. Professor Bell stated that White was tenderly attached in his younger days to a young lady, Miss Hester Mulso, the sister of an old college chum. Perhaps the truth will never be told concerning this matter. Unknown difficulties may perchance have stood in the way of a union; maybe that locked in Gilbert's heart there always burnt the holy flame; maybe that far from his being a life without romance, the whole of it was one lifelong romance. On the other hand, it may be that this alleged romance was purely fanciful, and rests on no real ground.

Miss Mulso was seven years his junior. She became acquainted during her life with numerous literary celebrities, amongst whom we find Richardson and Dr. Johnson, to whose *Rambler* she contributed when only 23 years of age, and she succeeded in winning for herself a reputation among the literary women of the day. She was best known as Mrs. Chapone, the name of her husband, a barrister, whom she married in 1760. She was, however, left a widow ten months after her marriage. She was the author of *Improvement of the Mind, addressed to a Young Lady*, and other works. She died in 1801.

White was never married. His nearest and dearest companions were the people amongst whom he worked, and those few friends in whose society and in whose correspondence he found sympathy with his predilections for Nature.

It is no easy matter to so transport ourselves back a hundred years and more, as to thoroughly grasp the appearance and surroundings of a man who then lived. If we consider for a moment, we shall see at once the great change that has come over the fashions of a man's dress since the eighteenth century. We must imagine White, as he rambled along the Lythe, or followed the scent-laden path to the Priory Farm, dressed in the old-fashioned costume of the period. No doubt he wore a clerical wig, whilst knee-breeches and buckles were a necessity to a man of his position. Many years ago an old woman was asked what she remembered of Gilbert White. Her recollection was of a man who used to walk about the hollow lanes "tap-tapping with his cane," and stopping now and again to brush the dust off his shoes with his handkerchief, careful man that he was. White was something of a geologist before geology was recognised as a science, and perhaps the "tapping" with the cane was an indication of his taste in this direction. On the other hand, taking into consideration the almost impassable nature of some of the lanes as they then were, the "tap-tapping" may have been a necessity in order to prevent the wear and tear and other inconveniences attached to the wearing of hose and knee-breeches.

One account of our author describes him as a "small" man, of spare form and upright carriage, and that he often rode a pony. In one of his letters he speaks of himself as "attended daily (for though not a sportsman, I still love a dog) by a beautiful spaniel, with long ears, and a spotted nose and legs."

Frank Buckland records in his edition of "Selborne"

(Macmillan), that a villager when questioned about Gilbert White, replied that he was thought little of until he was dead and gone, after which he was thought a great deal of. This is unfortunately true of too many people. A great man is almost without exception but half appreciated by t[illegible]neration in which he lives, whilst those whom [illegible]tion so oft delights to honour amongst its contemporaries, are just as often completely neglected or forgotten by its successors. Possibly not one in a hundred who know Gilbert White's life-work would have heard of Pennant's *Miscellanies,* had these not been referred to in the letters; perhaps even Pennant's name itself would have been quite unknown to them. Yet Pennant possessed a reputation in those days, which White probably never anticipated rivalling, much less excelling. And White's work has so far outlived that of Pennant, that it has taken a position in the classical world from which there is little possibility of it ever being removed.

In December 1830, an article appeared in *The New Monthly Magazine*, in which the writer related that an old dame who had nursed several of the family, told him that White "was a still, quiet body," and that "there wasn't a bit of harm in him, I'll assure you, sir, —there wasn't indeed."

In 1858, Robert Mudie wrote (in Blyth's edition) as follows: "There are still living in Selborne some old men who remember Mr White, and who speak of his memory in terms of the highest respect, as an excellent and amiable man. They do not appear to be very strongly impressed with any overshadowing greatness of character which he exhibited when he sojourned among them. They do speak, however, of his kindness and familiarity with his fellow-villagers,"

A villager who died in 1855, gave the following characteristic picture of White, viz., that he was a "little thin, prim, upright man." From what we have seen he may have been marked by small-pox. Add to this the information gleaned by Buckland from a woman who was eleven years old when Gilbert White died, that he was then a quiet old gentleman, who had many old-fashioned sayings, and gave away many presents to his parishioners at Christmas, and we may fairly claim to be able to form a striking mental picture of this great and good man. We are told, also, that the earnestness which he threw into his duties as a country clergyman, would scarcely be realised by one who expected to gain an appreciation of the man's character by the simple reading of his book. The use made of a sermon of his, printed in 1863 in the *Journal of Sacred Literature* and preached at different times on at least forty-two known occasions, tends to show the esteem in which his own original discourses were and have since been held. It may be imagined that in his day there were not a large number of curates who wrote, as well as preached, their sermons.

Professor Bell, who for years lived in White's own house, and who edited one of the best and most valuable editions of "Selborne" asserts that "His life may be truly said to have been passed in the constant and diligent fulfilment of the duties of his holy office, continuously from the time of his taking deacon's orders in 1747, till his death in 1793."

A letter reached me in 1902 from Mr. E. R. Small, in

which he said,—"My ancestors are old inhabitants of the village and neighbourhood, and my great grandmother, who lived to be 100, was the only remaining inhabitant who remembered Gilbert White when the village was visited by the naturalist, Frank Buckland."

Again and again his admirers have sought for some true picture of him, in order that the man whose moral and mental attributes are so well understood, might appear visibly clothed in the integuments of frail humanity. Yet so little he appears to have been thought of at the time, and so little considered as otherwise than a somewhat eccentric parson—probably there were many more in existence—that there was little or no demand for his portrait, nor any possibility of posterity ever demanding it. So we seemed to have been left in ignorance as to the features of the great naturalist.

It has been stated that in the large engraving of Selborne, published in the quarto edition of 1813, the figure on the brow of the hill is intended for Gilbert White. The late Earl of Stamford said that Dr. John T. White, a grandson of Benjamin, the publisher of this edition, stated that he distinctly remembered being told that the figures represented in the frontispiece were Thomas White and Mr. and Mrs. Yalden, with Gilbert on the brow of the hill, and that the latter was a very fair portrait of the general outline and appearance of the figure of Gilbert White.

As stated elsewhere, Professor Bell in his edition gives a different explanation of the four figures shown in the picture, and these do not include Gilbert White.

In *Notes and Queries*, No. 204, page 304, Mr. A. Holt White wrote as follows:—"Oriel College, of which

Gilbert White was for more than fifty years a Fellow, some years since offered to have a portrait painted of him for their Hall. An enquiry was then made of all the members of his family, but no portrait of any description could be found. I have heard my father say that Gilbert White was much pressed by his brother Thomas (my grandfather), to have his portrait painted, and that he talked of it, but it was never done." (This passage is quoted in Jardine's edition of 1853.)

Some mention should perhaps be made of the reputed portrait in oils of Gilbert White which was found in Caledonian Market a few years ago. There are, however, considerable elements of doubt as to its authenticity and it is not generally accepted.

A remarkable find was made in 1914 of a first edition of Pope's "Iliad," in six volumes, in the 3rd and 5th volumes of which are sketches of what I regard as authentic portraits of the naturalist. On the flyleaf of Volume I there is the following inscription in Gilbert White's handwriting: "Gilbert White, Oriel College, Oxon; given to me by Mr. Alexander Pope on my taking the degree of B.A., June 30th, 1743." On the flyleaf of Volume III is a pen-and-ink portrait with the inscription, "portrait of G. W., penned by T. C." On the fly-leaf of Volume V is a small pen-and ink portrait of what is undoubtedly the same face and head, and wearing a mortar-board. In both portraits are shown the long flowing hair reaching to the shoulders, with a delicate somewhat obtrusive nose and rather receding forehead. They both bear a distinct resemblance to the portrait formerly in the collection at Knebworth Hall, which bears an inscription that it is a portrait of

PEN-AND-INK SKETCHES OF GILBERT WHITE
*In his copy of Pope's "Iliad" in the British Museum.*

Gilbert White. I am of opinion that we have now good evidence of what the naturalist was like in his younger days. The six volumes of the "Iliad" have been acquired by the British Museum. The three portraits are here reproduced.

The character of the man—his exactness even to the most trivial details of his daily life, the trait of carefulness and precision which caused him to mark his sermons with the date of each occasion on which it was delivered—is nowhere more apparent than in his diaries. These extend over a period of nearly twenty-five years, dating from 1768 to 1793. They are in the possession of the British Museum. They were acquired by purchase in 1881 from the Rev. G. Taylor, who received them from a Mr. G. Soaper. Extracts were made by Edward Jesse (one of White's editors), and published in his *Gleanings in Natural History*. The last entry in the diary is dated June 22nd, 1793, that is to say, four days before that of his death. Here was vigour indeed in an old man. Right up to the last the same intense love for nature animated him, and found expression through the point of his pen. Spring had followed winter, the summer had succeeded spring, and still, as ever he had been, he was keenly on the alert to note the arrival of the birds of passage. Even as he had watched the movements of the fly-catchers for more than forty years, he was still just as keen in his observation of them a month before his death, as the entry for May 24th shews. Even as he had for year upon year watched the bursting of the buds and the opening of the flowers, so he still verified the entries in his naturalist's calendar, amending where necessary, and comparing dates with those of former years. Never does he seem to have tired of his duty, for a strict duty his hobby appears to have become. No word of impatience, no sighs or questioning as to the use or object of all his care and all his labour. His twenty-five years' record was made without thought of reward, and yet how much sweeter must have been the record of the last few years of his life, when the success of his book had been assured, and when he must have felt that meed of pleasurable reward which came from the knowledge that the results of his observations were not confined to himself, but that they would be scattered abroad for the good, the pleasure, and the benefit of a very wide world.

The neatness and exactness of the entries throughout the diaries themselves, speak eloquently of the manner of man he was. We can well understand that the man who wrote thus, would also be the man who would, as we are told he did, now and again stop in his walks "to brush the dust off his boots."

The feeling often comes over one when reading the life or the diaries of a great man, that there is something awe-inspiring in the fact that the reading of a few hours covers, it may be, a period of fifty, sixty, or seventy years. It seems almost irreverent thus so rapidly to pass over these sacred periods, when the man was forming, so to speak, his greatness, to hurry from one episode to another, from one trial to another, in a most trivial and careless manner, as though these episodes in the life had not been the real and tangible things that they were to the actor, as though one were reading some fine piece of fiction instead of the noble record of a real man's life. So it seems in reading this twenty-five years' diary, as we pass on from year to year; we see the winter disappearing before the ever-strengthening sun of spring; the migrants arrive, the migrants go; the hands pass round the floral clock as first the early form of vegetation, then the summer garb, and finally the autumn tints, all in turn show themselves, only to disappear before the oncoming of the succeeding season. So it goes on year by year, and as we read we seem to pass rapidly, hand in hand with the great naturalist, from one place to another, from one season to another, from one year to another, the observations of each fresh period amending or confirming those of a previous period. Surely such a man, if anyone, might rightly consider himself justified in theorising from so comprehensive a series of observations.

We find that the idea of keeping calendars of observations came from the Hon. Daines Barrington, who is termed the "inventer" of them. One of these calendars was annually presented to Gilbert White from 1768 onwards. We must bear specially in mind how important they became, at a time when so much natural history fiction, rather than fact, was prevalent. White made a fair start towards laying by that store of observation which is now admittedly so necessary to ensure rational deduction and theory, but which hitherto had not seemed so necessary to his contemporaries, many of whom repeated without question the extraordinary tales of the old classical naturalists. He may, in fact, be well called the "Father of British Popular Naturalists." In his lifetime he stood almost alone in the position he held; he has been followed, however, by a host of imitators, many of them most able men, and the army of present-day natural history writers are the direct literary descendants of Gilbert White.

In a lovely passage in his diary, under date March 10th, 1793, he writes:—"The sweet peal of bells at Farnham, heard up the vale of a still evening is a pleasing circumstance belonging to this situation, not only as occasioning agreeable associations in the mind, and remembrances of the days of my youth when I once resided in that town; but also by bringing to one's recollections many beautiful passages from the poets respecting this tuneable and manly amusement, for which this island is so remarkable."

This was but three months before his death, and there is a pleasing but painful interest in the passage. He had reached that age when man looks back with poetic sentiment upon the surroundings and associations of his youth. He must even then have felt that he had not many more months to spend amongst those that had known him so long. What must have been the peace of mind which such an one as he possessed, and how happy indeed must have been the final scene of all!

But a few months after the publication of his book, he wrote to the Rev. R. Churton, and expressed the pleasure which he felt in the words, "my book is still asked for in Fleet Street." What would be his feelings now, when his book in not only asked for in Fleet Street, but throughout the wide, wide world, from Great Britain to the Antipodes, from India, New Zealand, or South Africa; when it is found to have passed through one hundred and more editions, numbering an incalculable number of copies?

Settling down in 1755, in his native village, we find

that he spent thirty-eight years of peaceful quietude amongst the works of creation which he loved to contemplate. Here, indeed, is a picture to conjure up to one's mind. We who live amidst the turmoil and anxieties of the twentieth century, regard almost with envy, not so much the opportunities of peace and loveliness combined which he possessed, as the sweet docility of temperament which could so exist, without even a hankering after the excitement and wild dissipation of town life; which could remain for almost four decades content to do duty in the humble mode of life to which he had been called.

He died full of years, having passed the allotted span of life, little dreaming that the record of his serene and useful existence would bring him a degree of fame in the history of the literature of his native country, beside which many a more pronounced, many a more brazen fame of contemporaneous history would pale into the narrowest insignificance. His is the record of a quiet and blameless life, dedicated in its own humble way to the benefit of humanity: as such he will ever be regarded with a favour which increases with each generation, giving him a position from which it will be increasingly difficult for any detractor to dislodge him.

White was in many respects fortunate in regard to the place which knew him as the place of his birth, and in which he spent more than one half of his years. We dare not say that had he lived in a more uninteresting village, he would not have written as interesting a series of Letters as those which he has left us. A man of his disposition, in apparently the most unpromising parish would probably have found sufficient and more than sufficient material to have occupied his observant mind. Eyes that see are not always the eyes that observe. The mind's eye is not alone served by the material eye, and in order to produce similar results to those White left to us, each must be dependent upon, and in full accord with the other. White had the advantage of a most interesting parish, and he was undoubtedly the right man in the right place. His was a happy life made content by silent communings with Nature. The influences of a happy and contented life are never lost. Some of Gilbert White's facts may be now out of date, but his

SELBORNE STREET SHOWING "THE WAKES" WITH ITS OLD FRONT

book will ever live. In its influences lie its charms, and through it the happy and contented philosophic Gilbert lives. What though he lies beneath a simple gravestone in his native churchyard, marked only by a plain

"G.W."; his is a name which requires no monument. *Si monumentum quaeris,* his book supplies it.

His old friend, John Mulso, his colleague and companion at Oriel College, and brother to Mrs. Chapone, visited him on the 14th June, and left him on the 15th, as he tells us in the diary. Vigorous to the last, he at length slept, within two weeks of completing his seventy-fourth year, leaving for us a monument of patience and perseverance and an example of peacefulness and contentment.

## CHAPTER II

## HIS BOOK—"THE NATURAL HISTORY AND ANTIQUITIES OF SELBORNE"

FROM the lack of arrangement of the subjects dealt with in the letters of which the book consists, and the fact that many of them bear internal evidence of having been actually written in reply to others which the author had received, we may come to the safe conclusion, that at least the majority of them were not written with a view to publication, and that they have been presented to the world as they were written from friend to friend, with but little subsequent attempt at rearrangement or condensation.

The two series of letters were written to two gentlemen of similar pursuits and interests to those of the author. The first forty-four letters were addressed to Thomas Pennant, Esquire, but the first of them which bears a date is Letter X, viz., August 4th, 1767. The last of this series is dated November 30th, 1780, and they are all written from Selborne.

These are succeeded by sixty-six letters to the Hon. Daines Barrington, the first of which is dated June 30th, 1769. The 55th letter is dated October 10th, 1781, but the concluding eleven letters of this series bear no date at all, except the last, which is dated June 25th, 1787. Those that are dated were written from Selborne, with the following exceptions:—Letter VII is dated from Ringmer; Letter X from Fyfield; and Letter XVII from Ringmer.

These are followed by twenty-six letters on the "Antiquities of Selborne," bearing no particulars of time or place of writing. They were probably added with a view to the completion of White's "idea of a parochial history."

The greater part of the materials of the book, therefore, formed portions of private letters which Gilbert White wrote to his friends during a period between 1767 and 1781. He was forty-seven years old when he commenced to lay the visible foundations of that work which was afterwards to render his name famous, and he had consequently passed through many years of natural history experience, laying by the store of knowledge which so bountifully showed itself when he commenced his correspondence on natural history.

We are able to fix fairly well the time at which the idea of publishing a work on natural history began to take shape in his mind. In his 5th letter to Barrington (1770), written scarcely a year after the commencement of his correspondence with that naturalist, but when about three-fourths of his correspondence with Pennant had been completed, he writes, "When we meet I shall be glad to have some conversation with you concerning the proposal you make of my drawing up an account of the animals in this neighbourhood.

"Your partiality towards my small abilities persuades you, I fear, that I am able to do more than is in my power: for it is no small undertaking for a man unsupported, and alone, to begin a natural history from his own autopsia! Though there is endless room for observation in the field of nature, which is boundless, yet investigation (where a man endeavours to be sure of his facts) can make but slow progress: and all that one could collect in many years would go into a very narrow compass."

On the other hand, we cannot safely judge from this that he had then the intention of collecting his correspondence with a view to publication. Possibly he may have considered it as too trivial to take rank beside the more systematic treatises of his contemporaries. In the next year, however, as we learn from a letter to Pennant, he writes, "As to any publication of my own, I look at it with great diffidence, finding that I ought to have begun it twenty years ago." He was now fifty-one years old. "But if I was to attempt anything it should be something of a Natural History of my native parish, an '*Annus Historico-Naturalis,*' comprising a journal for one whole year, and illustrated with large notes and observations."

The naturalist's calendar was destined, however, never to appear in White's lifetime. In 1771, he writes to his brother John again on the subject, and says, "As matter flows in upon me, I begin to think of composing a Natural History of Selborne in the form of a journal for 1769. We shall then be able to compare the climates." On March 29th, 1774, he again writes to John, "As to my letters, they lie in my cupboard very snug: if you will correct them, and assist in the arrangement of my journal, I will publish. I have finished the monography of all the *Hirundines* except the swift."

Again, on April 29th, 1774, he writes to John as follows:—

"Out of all my journals, I think I might collect matter enough, and such a series of incidents as might pretty well comprehend the nat. history of this district, especially as to the ornithological part, and I have moreover, half a century of letters on the same subject, most of them very long; all which together (were they thought worthy to be seen) might make up a moderate volume. To these might be added some circumstances of the country, its most curious plants, its few antiquities, all of which together might soon be moulded into a work, had I resolution and spirits enough to set about it." (This letter and about 90 others were sold at Sotheby's in April 1905. All were addressed to various near relatives.)

Fifteen years more were, however, to elapse ere White's book was ready for publication.

In March, 1775 he wrote;—"I have had an heat and stiffness in my eyes from overmuch reading, that made writing very irksome for some time."

On Feb. 17th, 1776, he wrote to Samuel Barker;—"Mr. Barr[ington] wants me to join with him in a Nat. Hist. publication; but if I publish at all, I shall come forth by myself. Bro. Tho. is laying up materials for a History of Hants. He will some day take an artist down."

On Nov. 23rd, 1780, he wrote to Barker;—"You do not, I hope, flatter me about my Nat. Hist.; if you do not, I am much pleased to find that an intelligent person like yourself approves of it. Were it not for the want of a good amanuensis, I think I should make more progress; but much writing and transcribing always hurts me."

The difficulty which he experienced in writing receives a sidelight in a passage in a letter to John White in May 1776 (?). He says;—

"My left hand is full of gout; all my fingers look red

and shoot and burn. If I have gout about me, it is best to come out."

On Jan. 8th., 1788, he writes to Barker:—

"I have been very busy of late, and have at length put my last hand to my Nat. Hist. and Antiquities of this parish. However, I am still employed in making an Index—an occupation full as entertaining as that of darning stockings, tho' by no means so advantageous to society. My work will be well got up, with a good type and on good paper, and will be embellished with several engravings. It has been in the press some time, and is to come out in the spring."

When, finally the book was published, White's ideas had considerably enlarged, for it was found that in his preface he had stated that he wished to "lay before the public his idea of a parochial history, which ought to consist of natural productions and occurrences as well as antiquities." He was wise in thinking that such things could not be more attractively set out than in the letters which he had for years been inditing to his friends, but of this prior idea of the Naturalist's Calendar, we obtain just a glimpse in the last letter to Barrington, for he there concludes with these words:—

"When I first took the present work in hand I proposed to have added an '*Annus Historico-naturalis*, or the Natural History of the Twelve Months of the Year'; which would have comprised many incidents and occurrences that have not fallen in my way to be mentioned in my series of letters; but, as Mr. Aikin of Warrington has lately published somewhat of this sort, and as the length of my correspondence has sufficiently put your patience to the test, I shall here take a respectful leave of you and natural history together, and am,

With all due deference and regard,

Your most obliged and most humble servant,

GIL. WHITE

SELBORNE, *June 25th*, 1787."

Fortunately for posterity his manuscripts at his death fell into the careful hands of his brother Benjamin White, the bookseller and publisher of Fleet Street, with the result that in 1795, Dr. Aikin, with whom he apparently shrank from entering into competition, prepared and extracted therefrom the Naturalist's Calendar which White had prepared, and added thereto from White's papers a selection of "Observations on Various Parts of Nature." These have appeared regularly with almost every succeeding edition of his book which has been published.

Although Gilbert White issued but one book, we must be careful to remember that the matter contained in it was the work of a lifetime. We may not agree with all his conclusions; his facts may not always have been correct, and no doubt many subsequent natural history works have long superseded his, in so far as they are standard works of reference; but the charm which is found in this book is found in none other. It is an innocent peaceful picture of the life of an unassuming country parson; who recorded what he saw, and who in every record which he made, stamped it with his own blameless character. White's soul breathed into his book, and in it he has left to us a veritable part of himself.

He did not write for the world to see. The laboriousness with which he wrote his lengthy letters to his friends, and embodied in them the results of his observations, at a time when letter-writing was a greater art and certainly was a greater labour than now, and when the conveyance of such letters for long distances must have been a matter of considerable expense,* shows well that he could have had no other motive in writing, than that furnished by his love to dwell upon the creatures that interested him, and by his willingness, at the suggestion of his correspondents, to place on record the results of his own careful observations.

Probably when finally publication was decided upon he wrote the nine introductory chapters (which bear no date) and perhaps the last, with the deliberate intention of rounding the series off, and completing the elements of the village history, but most of them bear the genuine stamp of the familiar letter from friend to friend, and were evidently in the first place never intended for the public gaze.

The final taking leave of his correspondent "and of natural history together" always seems to strike me with a view of sadness, especially as by that time he had in all probability really taken leave of most of his extensive letter-writing. The last letter is dated June 25th, 1787, and when in 1789 his book appeared, he had already entered his threescore-and-tenth year, and its publication must have seemed to him almost as the closing of the volume of his life.

"Your work," wrote John Mulso, "upon the whole, will immortalize your place of abode as well as yourself," and this prophecy has been wonderfully fulfilled.

*We know, however, that he used 'franks' when he could obtain them.

The two series of letters, although to some extent concurrent as to time, were kept separate when first published. Some of them had become a good deal fingered and creased, and this necessitated the making of fair copies for the printer. This was done for the most part by White himself, but a few were copied for him by an amanuensis; a nephew, probably one of the Barkers,* was thus employed by him. These copies, 106 in number, were sold at Sotheby's on April 26, 1895, and went it was stated to America. About three-fourths of the actual letters themselves (1767–1773) are now in the Bitish Museum having been purchased for £134 in 1897. In the catalogue of manuscripts they are numbered 35138. (See pp. 70–71.)

The naturalist's dread of the reviewers was to some extent removed by his brother's undertaking to give a review of the work in the pages of the *Gentlemans' Magazine*. This appeared in the year 1789. The following is an extract from it.

"Contemplative persons see with regret the country more and more deserted every day, as they know that every well-regulated family of property, which quits a village to reside in a town, injures the place that is forsaken in many material circumstances. It is with pleasure therefore, we observe, that so rational an employment of leisure time as the study of nature promises to be popular; since whatever adds to the number of rural amusements, and consequently counteracts the allurements of the metropolis, is, on this consideration, of national importance.

*Gilbert's sister, Anne, married Thos. Barker of Lyndon Hall, Rutland.

"Most of the local histories which have fallen into our hands have been taken up with descriptions of the vestiges of ancient art and industry, while natural observations have been too much neglected. But we agree with Mr. White in his idea of parochial history, which, he thinks, ought to consist of natural productions and occurrences, as well as antiquities: for antiquities, when once surveyed, seldom recall further attention, and are confined to one spot; whereas the pleasures of the naturalist continue through the year, return with unabated attractions every spring, and may be extended over the kingdom.

"Mr. White is the gentleman who some years ago favoured the world with a monography of the British Hirundines, published in the *Philosophical Transactions*, which we reviewed in a former volume. It is now reprinted, and the same sagacity of observation runs through the work before us.

"On the whole, we will pronounce that the enquirers into natural knowledge will find Mr. White to be no unequal successor of Ray and Derham; and that the History of the Priory is a curious tract of local antiquity. We should not hesitate to speak so favourably of this work, even though it had much less rural anecdote and literary allusion to recommend it."

Jesse states (1851) that not long after publication, the following remark was made to a nephew of White by Dr. Scrope Beardmore, the Warden of Merton College. Jesse received the ancedote from the said nephew. "Your uncle," the Warden said, "has sent into the world a publication with nothing to call attention to it but an advertisement or two in the newspapers; but depend upon it the time will come when very few who buy books will be without it." A far-seeing gentleman was the worthy Doctor.

Gilbert White was a fervent Linnean. In a paper which appeared in a Swedish publication in 1930 White's influence is discussed with leading naturalists of the day. In reviewing this paper Dr. Knut Hagberg shows that Linnean nature-study took a firmer hold in England than in Sweden, and states that through White's work it influenced the poetry of the time, in Keats, Coleridge and Wordsworth.

This chapter may well be concluded with the rhyme that he addressed to himself, which I think shows the temerity with which he became an author:–

"Go, view that house, amidst the garden's bound,
Where tattered volumes strew the learned ground.
Where Novels, Sermons, in confusion lie,
Law, Ethics, Physics, school-divinity;
Yet did each author, with a parent's joy,
Survey the growing beauties of his boy,
Upon his new-born babe did fondly look,
And deem Eternity should claim his book.
Taste ever shifts; in half a score of years
A changeful public may alarm thy fears;
Who now reads Cowley? The sad doom await,
Since such as these are now may be thy fate.

GIL. WHITE

South Lambeth, 25 June, 1789."

## CHAPTER III

## THE VILLAGE OF SELBORNE

A DESCRIPTION of the village of Selborne as it appeared in White's time would not greatly differ from that of the village as it appears to-day. There have been, of course, alterations in some respects, but the general aspect of the village and of its approaches, is much the same now as that which it presented in the days of the great naturalist.*

As approach is made from the town of Alton, we pass along a high road which soon presents us with a view in the distance of the Hanger Hill, with Nore Hill hidden, so to speak, almost behind it. Before any part of the village of Selborne is at all visible, the Hanger forces itself on the eye with the involuntary exclamation, "this must be the far-famed Hanger;" and so it is.

In passing along the present highway to Selborne, the hills are visible for at least four of the five miles which the pilgrim must traverse.

About a mile from the parish church we pass Norton Farm, where a road on the right takes the traveller to Faringdon. Here also the old lane branches off, along which all traffic formerly passed to Selborne before the road which we are traversing was made. It is a good example of the hollow lanes to which White gives mention in his fifth letter to Pennant:–

"Among the singularities of this place the two rocky hollow lanes, the one to Alton, and the other to the forest, deserve our attention. These roads, running

*This was written in 1896. Unfortunately there have been changes since. Many of the picturesque but timeworn cottages have given place to modern structures.

through the malm lands, are, by the traffic of ages, and the fretting of water, worn down through the first stratum of our freestone, and partly through the second so that they look more like watercourses than roads; and are bedded with naked rag for furlongs together."

This lane, bad enough in his time for travellers, had become, by disuse, almost impassable for vehicles of any large size, but, as he also remarks, they delight the naturalist, owing to the shelter which they give to all manner of flowers and ferns in their nooks and crannies. This hollow lane enters the village by Gracious Street, and then curves round to the Plestor, to which we shall refer later.

Passing on from Norton to Selborne we find that the road commences to slope downward, and we may be sure that when we reach the bottom we shall find a brook or stream of some dimensions. In fact we here find the first of the two streams, one of which flows across each end of the village. The village street runs, roughly, from N.N.W. to S.S.E.:—

"At each end of the village, which runs from south-east to north-west, arises a small rivulet: that at the north-west end frequently fails; but the other is a fine perennial spring, little influenced by drought or wet seasons, called Well-head."

The one which is first encountered is that which on certain occasions fails.

Before we cross over the stream, it is worth while noticing that as we enter the village, the church and other buildings stand on the rising ground beyond the stream, and extend lengthwise to our left before our view, when not hidden by the density of the foliage.

On our right is a charmingly-situated house, the meadow extending therefrom down to the stream, whilst opposite to it on the other side is a public foot-path, passing by numerous stiles into the road which is a continuation of the old rocky lane to Alton.

We pass over the stream by a bridge which stands at some height above the water, and which removes the necessity of too great a pull for vehicles up the road on the other side. This bridge is, however, of modern construction, having been built some sixty-five years ago. In White's time there was here no bridge for vehicles at all, nor even a road, a little footbridge which we can see below us, seemingly placed loosely together, in the form of a Norman arch, being sufficient for the foot-traffic of the day. There were also some foot-bridges below the church and vicarage.

As we mount the road, we see a portion of the Long Lithe on our left, whilst almost towering above it are the walls of some gardens, which present to the pedestrian an appearance as of a fortified entrance to the village. A little higher and we are on level ground. Here the road is joined by the old Alton road, this part being still in constant use, as it forms the first portion of the road round the hop plantations, and over the hill, to Newton.

In Letter V to Pennant previously referred to, White speaks of the hollow lanes being caused by the "fretting of ages" and at the junction of these roads we shall notice on our right that the garden at the corner, in front of the cottage, is situated at the level of the top of a wall, about four-and-a-half feet in height.* If we pass

*The cottage has since been pulled down.

down the Newton road, the gardens on our right are more or less similarly situated, whilst on the left is the high bank through which the road has been cut.

We will now resume our route on the main road. As we turn slightly to the left we see at once the square plot of ground known as the "Playstow," a name spelt in many different ways, which constituted the village green:—

"In the centre of the village, and near the church, is a square piece of ground surrounded by houses, and vulgarly called the Plestor. In the midst of this spot stood in old times, a vast oak, with a short squat body, and huge horizontal arms extending almost to the extremity of the area. This venerable tree, surrounded with stone steps, and seats above them, was the delight of old and young, and a place of much resort in summer evenings; where the former sat in grave debate, while the latter frolicked and danced before them. Long might it have stood, had not the amazing tempest in 1703 overturned it at once, to the infinite regret of the inhabitants, and the vicar, who bestowed several pounds in setting it in its place again: but all his care could not avail; the tree sprouted for a time, then withered and died. This oak I mention to show to what a bulk planted oaks also may arrive."

The storm referred to was one of some notoriety, and an account of it written by Defoe is extant. It was in this storm that Winstanley lost his life in the destruction of Eddystone Lighthouse, when twelve British men-of-war were lost off our shores, and over 1,800 men perished with them; when it was estimated that over 8,000 persons perished through floods and devastation around the coast, and when damage was done in

London alone to an estimated amount of two millions sterling.

On the left hand side of the Plestor stands the Vicarage, whilst at the head of the Plestor is the gate leading to the graveyard and the church.

Immediately inside the railings of this quiet little cemetery is a very fine yew tree. Yew trees have long been held to be the most suitable denizens of country churchyards. This one must be of great age. Thomas Barker, Gilbert's brother-in-law, has a record of it in one of his note-books, now in the possession of the Rev. F. Gilbert White, of Leusdon Vicarage. It runs as follows:—

"May 5, 1750. The yew tree in Selborne churchyard—

| | |
|---|---|
| Girth at the root | 18ft. 11in. |
| Girth at middle of body | 22ft. 6in. |
| Girth at height of body | 6ft. |
| Topmost bough about | 35ft." |

When White wrote Letter V of the "Antiquities," he recorded its girth as upwards of 23ft. This was possibly thirty five years later.

In July, 1912, Mr. Hugh Boyd Watt stated that the yew measured 25ft. 5½in. circumference at the narrowest part of the trunk accessible, and it was about 68 feet in height.

All around its base is a narrow seat, where the weary traveller may obtain a welcome, although perhaps but a scanty rest.

Beyond the yew tree stands the western end of the church. In Letter IV to Pennant is the following passage:—

SELBORNE CHURCH FROM THE SOUTH-EAST
*showing the Porch and Tower, and the Yew Tree.*

"From a notion of rendering their work the more elegant, and giving it a finish, masons chip this sandstone into small fragments about the size of the head of a large nail, and then stick the pieces into the wet mortar along the joints of their freestone walls; this embellishment carries an odd appearance, and has occasioned strangers sometimes to ask us pleasantly, 'whether we fastened our walls together with tenpenny nails.'"

GARNETTED WALL OF CHURCH, STUDDED WITH "TENPENNY NAILS."

If we look straight ahead we shall see that the nearest portion of the church is similarly studded with these tenpenny-nail-like pieces of ferruginous sandstone.*

*In some of the walls around Windsor Castle, as well as in West Sussex, the same practice is apparent, except that here flakes of flint are used instead of sandstone. I am told that it is called "garnetting" or "garnet-jointing," and that it is used to prevent the ravages of frost on the mortar.

If we walk round the church to the right we immediately pass by the southern porch. If we turn to the left, we pass the belfry tower, but whichever way we may go, the point to which all pilgrims wend their steps with reverend foot, is the humble tombstone faintly marked with lichen-covered letters "G.W."—just

GILBERT WHITE'S GRAVESTONE.

plain unassuming "G.W."—standing near the north-eastern corner of the church in a row of half-a-dozen other stones of similar diminutive size. The Rev. Ch. Taylor, the Vicar of Selborne, who registered Gilbert White's death in the church records, little thought that when relegating the mortal remains of his curate to the most obscure corner of God's acre, that that corner would be the one to which people would travel from all corners of the universe, to bear the testimony of their admiration to the illustrious dead.

A large modern vault close by contains the remains of one of White's admirers and editors, viz., Professor Thomas Bell, F.R.S., who died in 1880.

Looking due east we see the gradual sloping of the ground towards the Lithes, down to the peaceful bottom where the waters meet, and where the single stream pleasantly meanders amongst the rich meadows between the hills.

Mr. T. W. Shore, the Hampshire antiquarian, said that "the church is of Norman foundation. The manor was an ancient domain of the Saxon Kings, and was held of them by the Bishops of Exeter, with whom it was connected for many centuries. In Domesday it is said to have been held by the king, and belonged to Bosham, where there was a religious house."

Gilbert White only carried the existence of the church back to the time of Henry VII. It is considered, however, by archaeologists that the style of the church is that of the Transitional Norman period. The chancel has three Early English lancet windows. The south aisle terminates in a chantry, which was probably founded at the same time as the church. Restoration was rendered necessary in 1883, owing to the collapse of the walls. As one sits in the ancient church facing the east, one is struck even now by the fact that the massive Norman pillars on each side of the central aisle appear to be falling outward to a serious extent. A buttress has been built outside in recent times to counteract the tre-
d

mendous outward pressure of the roof, near to the spot of White's gravestone. It is on this side on which the outward lean is most noticeable.*

In front of the south aisle are two stone coffins which were disinterred by Professor Bell, and in which were found two skeletons. These two coffins are additional confirmation of the age of the church.

Other stone coffins which have been more recently discovered beneath the foundations of the church itself, are there to be seen. I was informed by an eye-witness that in one of them a complete skeleton with a very slight amount of dust was found. The first act of the labourers who dug up the coffin was to ransack it well, in the expectation of finding treasure. The bones were afterwards burned.

Gilbert White's little gravestone lies fifth from the north wall of the aisle, and bears the inscription: "G. W., 26th June, 1793."

The memorial slab to White is, curiously enough, on the south side of the chancel, although it is stated thereon that the great naturalist and observer lies buried beneath the grave marked by the fifth stone "from this wall." The wall, however, to which it is now affixed, is the one farthest from the grave, and this mistake, it is to be feared, must have created some uncertainty in the minds of many strangers as to the exact position of the grave.

Over the altar is the triptych (see page 104) which was

*The church does not exactly correspond with the cardinal points of the compass. The workmen are said to have been employed during the longest days, and so tried to set the church pointing to the rising sun.

TREES PLANTED BY GILBERT WHITE
*in front of the old butcher's shop.*

presented to this church by Benjamin White, the brother of Gilbert, and the original publisher of his book. It represents in the centre picture the Adoration of the Magi, and is painted on wood. Some years since it was found in a dilapidated condition in the vestry, and, owing to the interest roused in it by some correspondence which appeared in *The Field,* a county subscription for the restoration of the panels and their replacement in the church was set on foot, with the result that the triptych was restored and placed in its present position.

Passing, in continuation of our description of the village, round the churchyard, we reach the old yew tree, and make our exit by the gate which leads us again to the Plestor. There are some fine, comfortable-looking old cottages on our left, which we pass before reaching the village street.

On gaining the main road or street, the distance is but a few yards onwards to "The Wakes," Gilbert White's famous old house, which is found on the south-west side of the street. Here is the centre, the all-attractive nucleus, redolent with the best traditions of all that pertains to the sacred study of popular natural history. Here we pause awhile, to gaze in silence on the time-honoured homestead, ere we pass on to view more of the village which witnessed the labours and the loves of the great man.

We propose to devote a short chapter specially to the house and its immediate surroundings. Immediately opposite the house is what was once the butcher's shop, in front of which are three of the four limetrees which Gilbert planted in 1756 "to hide the sight of blood and filth." The cottage next to it was burnt down, and a modern house erected.

Still a few yards on, a meandering road leaves us on our left, which leads to Dorton, lying below the Lithes, and this finally conducts us to the Priory Farm. In the materials of which the farm-house is built are found all that remains of the old religious house, with the exception of numerous encaustic tiles strewn about in careless negligence of their archaeological value, and a large stone sarcophagus resting in the meadow. There is a small portion of a thick wall still standing, which, it is thought, is the only piece of the original structure remaining. This is at the farthest corner of the farm-house garden, and the roof of a tumble-down summer-house formerly rested upon it. The fine old tiles were trodden under-foot with impunity in the summer-house, whilst exposure to the atmosphere, the effects of rain and frost caused them irreparable mischief. The inlaid coloured portions were separating from the matrix, and their patterns were being obscured. The authorities of Magdalen College, Oxford, are now fully alive to the importance of their neglected heritage.

The patterns of the tiles were very various. Among them were representations of fish, others were of birds, together with many symmetrical designs, fleurs-de-lis, &c. A few of them have been figured in Buckland's edition of Selborne. Some bear architectural designs, and it may be that they portray certain features of the Priory.

The road between the Priory Farm and the village offers fine scope for the botanist, and for picturesqueness is scarcely surpassed in the neighbourhood. In a

ENCAUSTIC TILES FROM THE PRIORY SHOWING ARCHITECTURAL DETAILS.

copy of Hudson's *Flora Anglica* (1762) which White used, 439 species of plants are enumerated as having been found at Selborne. It can well be imagined that this was a favourite hunting ground of his.

From the point in the main street where this road commenced, the cottages were in White's time almost all on the right-hand or west side, *i.e.*, the same side on which is "The Wakes." On the opposite side was that fertile tract of black vegetable soil, which White mentions in his last letter to Pennant. This, however, has since been partially built upon, although the productiveness of the soil on this side of the road is still held to be altogether superior to that immediately beneath the Hanger.

The Hanger Hill approaches closely to our right hand as we pass along, and is soon separated from the road only by a short path which leads to the foot of the famous Zigzag.

It was in a cottage opposite to the Zigzag that I sojourned in 1902, when I was joined there by W. H. Hudson, who was then engaged in writing his *Hampshire Days*. I was able to show him where to get some grasshoppers that he needed, which he illustrated in the book quoted. The Zigzag is attributed in the village to the Danes, although we can scarcely now understand why it is so. There it is, and there it will remain, "and so 't will be when I am gone." The Zigzag consists of a path up one of the steepest parts of the Hanger, and is overhung by trees on both sides, but instead of being cut in one direct line, is excavated in zigzag fashion, in order to reduce the gradient of the path as much as possible. Fourteen times the path turns to the left, and fourteen times it turns

THE WISHING STONE AT THE TOP OF THE ZIGZAG

to the right. Each turn is about four times the distance of the equivalent vertical progress, so that instead of climbing upward rather steeply for about 200 feet, a distance some 800 feet in length is traversed at a somewhat gentler gradient. Perhaps the former is really preferable.

In rainy weather, the water rushes down the Zigzag as a miniature torrent, and the clayey chalk becomes extremely slippery, and, were it not that sundry bushes and young trees between the various turns of the path afford something to grasp, it would be impassable and unclimbable in such weather.

At the top of the path is a large block of stone, known as the Wishing Stone. It is a sarsen-stone. What could have been the object of bringing it there? Was it ever an altar-stone? Why was the Zigzag cut? There is a road but a short distance beyond, which encircling some cottages on the brow of the hill, gives far easier access to the sheep-down and common above.*

Close to the Wishing Stone is the Memorial Seat provided and dedicated by the Gilbert White Fellowship in May 1924.

At the foot of the Zigzag there is a private gate leading into Gilbert White's little Park, and here another path starts in a diagonal direction up through the Hanger, making its exit on to the common above at some distance to the north of the Zigzag. So far as I have been

*It is stated in the *Life and Letters of Gilbert White* that the Zigzag was made by White and his friends. Possibly it was remade and re-cut by them, but I am inclined to think that it was there long before White's time, and that the path which White cut is the improved zigzag which bends to the right at the foot of the ancient Zigzag.

able to ascertain, the old Hermit "hung his straw-clad cell" near the top of this diagonal path.

A good engraving of the Hermitage appears as a frontispiece to Vol.I. of Markwick's edition of "Selborne" (1802). It is satisfactory to see that even the Hermit was a good Catholic, since he is represented with his cross and rosary at his waist, whilst even at the apex of his strawed roof is a rude wooden cross. It was upon this wooden cross that the goat-sucker perched, which enabled White to make the observation that the jarring note of the bird was sufficient to set the whole structure in vibration.

Half-way up the Zigzag a path on the left leads to a small private plantation. Here was a favourite spot to which the late Professor Bell loved to retire, and where much of his study was carried on.

The three-cornered piece of land at the base of the hill is devoted to allotments. Skirting these, and taking a compass of the south-westerly termination of the Hanger, a road leads past some picturesquely situated "hanging" cottages, and bends backward as it rises to a cottage very close to the top of the Zigzag, but the road very quickly narrows as it passes the hanging cottages into a steep, and, at times, slippery footpath. Half-way up this hill the reservoir has been made, and to this further reference will be made presently.

We will, now, return to the main road. After passing the path leading to the Zigzag, there are several clean and tidy-looking cottages on the right, all built of the "freestone" in which the neighbourhood so plentifully abounds.

As we pass the former 'White Hart' Inn on our left,

it is worthy of notice that the gardens in front of the cottages on the opposite side of the road are at a height of between four and five feet above the level of the road, a circumstance which we noticed in regard to those at the other end of the village. The cottages now become few and far between, and the road divides into two branches, which almost immediately again unite,

COTTAGES BORDERING ON THE WELL-HEAD STREAM.

leaving an island, so to speak, on which are some farm buildings. The left hand branch crosses, by a little bridge, over the stream issuing from the Well-head, to which we here have our first introduction, and after sending off a branch-road, a hollow lane to Oakhanger dwindles into a footpath in front of some cottages, and following the side of the stream, again rejoins the main road. At the commencement of the road to Oakhanger (Oakganger, as it seems to be generally called) it is noticeable that the substantial-looking cottages which there appear, are really built on the site of an old quarry. Possibly the cottages consist of the stone of this very quarry in a rearranged condition. One cannot help pausing to note the picturesque situation of the cottages opposite the footpath, with the gentle rippling stream passing merrily along in front of their little cottage gardens.

WHERE THE WELL-HEAD STREAM CROSSES THE ROAD.

A few more steps, and a road leaves us on the right, up Galley Hill, betweenthe Hanger Hill and Nore Hill, and on to Newton. On the banks at the side of this road were found some of the largest wild bell-flowers which it has been my lot to meet. A little distance up the road is a pit or quarry, and by mounting this—for even this road is a "hollow lane"—we are enabled to look directly down on the cluster of trees and other foliage which

THATCHED COTTAGES IN SELBORNE STREET

mark the outcrop of the far-famed "Well-head;" to note the miniature valley which marks the path of the stream which issues from the spring, and to follow it with the eye until it makes its appearance at the side of the main road from which we have temporarily diverted our footsteps. We regain the road again, and find that a Mr. Mills, in years gone by, has faced the bank with an iron facing, causing the water to issue from a lion's mouth into a useful but small trough. On each side of the lion's head, the name of the benefactor was perpetuated by two windmills cast in relief on the iron face.

There are cottages, with their gardens on the high bank at the side of the road, opposite to the spot where the water issues through the cast-iron facing, but the village proper we have now left behind, and the road hence is confined between two high banks, which are covered with verdure at the top, the slopes constituting a rich carpeting of beautiful ferns and flowers. The road then leads past the foot of Nore Hill, the second of the twin hills which Gilbert White has rendered immortal.

Following on the centenary celebration of the death of Gilbert White, it was suggested by many admirers that it would be a fitting memorial to his greatness to bring the water-supply from the Well-head, which is situated somewhat beyond the village, up to the very doors of the villagers. The idea was taken up with energy, and sufficient funds were soon raised to allow of the project being successfully carried out. The water is forced by a hydraulic ram to the reservoir to which reference has been made, and the reservoir is at a sufficient height above the village to ensure the necessary pressure being obtained from the natural fall of the water. This is then conveyed to stand-pipes at various points in the village. It appears necessary to fill the reservoir only every third day, so that little, if any, alteration is made in the flow of water into the Bourne, as the stream is called. The reservoir is covered in, and the excavations made do not impair the beauty of the spot. It would have been a thousand pities if the natural beauty of this historic spot had received a blemish, even for so laudable an object. The supply is a tremendous boon to the villagers, especially to the old and feeble folk who had hitherto to carry it by hand from the Lion's-head to their cottages in some of the more remote parts of the village.

In connection with this important and useful work, a passage which occurs in White's diary is of extreme interest. He had been visiting the church at Bentley, whither he had walked with his brother, the distance being, as he states, one mile from his house. In noting this in his diary, under date March 15th, 1793, only three months before his death, he says:—

"The whole district abounds with springs. The largest spring on my brother's farm issues out from a bank in the meadow, just below the terrace. Somebody formerly was pleased with this fountain, and has, at no small expense, bestowed a facing of Portland stone, with an arch and a pipe, through which the water falls into a stone basin, in a perennial stream. By means of a wooden trough this spring waters some parts of the circumjacent slope. It is not so copious as Well-head."

From the top of the Zigzag to a distance corresponding to the whole length of the village road, stretches

VIEW FROM THE HANGER
*showing the Church in the middle distance.*

the Common, dotted here and there with clumps of trees, and in some places with tangled, verdant undergrowth, whilst the more open spaces afford sheepwalks for numerous flocks of sheep. This common, at the very summit of the Hanger, is indeed a lovely expanse to wander over. Sometimes the ground is covered with quantities of wild but luscious strawberries; ahead one sees beautiful prospects formed by clumps of wild rose bushes, whilst nearer are stretches of bracken, or mossy carpets figured into pattern by the little blossom of the wild thyme. Ever and anon the eye breaks in upon a perfect forest of gigantic foxgloves, whose crimson gives a most artistic setting to the sober green of youthful bracken. Oh, for an artist's pen to describe the many changing scenes!

It is not easy to keep one's bearings on this common. If you think you are making straight for the Zigzag, possibly you find you are just about to descend a precipitous incline to some hop plantations at the north end of the village. When you think to take a view in the direction of Faringdon and so go ahead to a clump of trees which you are quite sure that you recognise, you are surprised, as your view becomes clearer, to see a thickly-wooded hill in the distance, which looks remarkably like Nore Hill, and not until you see below the little valley of the Well-head are you convinced that you have been walking in exactly the opposite direction to that which you had anticipated.

At the north end of the common are three more or less narrow and long hollows which appear to have been scooped in remarkable manner out of the solid hill, and the approach to these is consequently on their south sides somewhat precipitous. Their formation is very remarkable, and is certainly due to the action of water, in some form. They are used for the purpose of hop and fir plantations. A road from the common skirts them, and joins the road from Newton to Selborne at the spot where is situated the great chalk-pit, to which White refers in his Letter III to Pennant. In his time it appears to have yielded numerous fossils.

Gilbert White's references to the geological characters of the neighbourhood are confined to his first five chapters. It will be necessary, in concluding this chapter to make some reference to the geology of the neighbourhood, since without exaggeration, it may be said, that it is attributable to the rocks and soils of Selborne, more than to any other characteristic that could be mentioned, that the natural history of the neighbourhood has been such as to attract the attention of the whole English-speaking world to the village, through the medium of White's book.

There is a wide tract of land known as the Weald, which is bounded on the north by the Surrey Downs, and on the south by the Sussex Downs. On the east and south-east it touches the sea coast, and on the west it reaches to the chalk downs which are a continuation of the great Salisbury Plain. At the first step eastward off the chalk lies the village of Selborne, placed therefore at the western extremity of the Weald.

It is thus situated in the proximity of the outcrops of some prominent and well-recognised geological formations. All around the Weald, the same formations exhibit themselves whenever a descent is made off the chalk hills. First there is the outcrop of the Upper Greensand, or "chloritic" series. Then comes the wider and more noticeable band of Gault Clay. This is followed by, as a rule, a still greater width of Lower Greensand, or neocomian strata, which in some parts rises to a height equalling that of the Downs themselves. Then from beneath the Lower Greensand rise in succession the Weald Clay, and the assemblage of strata known as the Hastings Sands. Horsham is the nearest approach of the Hastings Sands, and Haslemere that of Weald Clay to Selborne. The others are found within the parish itself.

Starting from the Hanger and Nore Hill, where we leave the Chalk and the Chalk Marl, and proceeding by the road down the Lithe to Priory Farm, we pass over the Upper Greensand until we reach a point about half-way between Selborne and the Priory. Here we tread over a tongue of the Gault Clay whose early outcrop has probably here been brought to light by the denuding action of the little stream, when it extended its volume higher up the sides of the valley. The line of outcrop of the Gault runs in a general north and south direction through the site of the Priory. Continuing eastward over the Gault fields, we shortly reach Oakwood, and Oakhanger Pond. As we reach them we have just stepped upon the outcrop of the Lower Greensand, a member of the Cretaceous system of strata.

As in the case of the Gault, the outcrops of all the rest run in a north to south direction. Of the places best known as being mentioned in the "Natural History," East Worldham, Harteley, Selborne village, Temple, Bradshott, and Empshott, are all upon the Upper Greensand. The Priory is upon the Gault, whilst Kingsley, Oakhanger, Oakwood, and Greatham, are situated on the Lower Greensand, as also is Wolmer Forest.

The excavations which were made in the Chalk Marl of Selborne Hanger, when the reservoir was made in 1894, afforded a good opportunity of inspecting the fossils of the formation. The usual chalk fossils were met with, although in a fragmentary condition. Portions of *Inocerami* were found in plenty, a few *Rhynconellæ,* with an occasional *Pecten quinque-costatus.* Sea-urchins, *Ananchytes* (*Echinocorys*) and *Micraster* were of course there: also one or two *Belemnites,* and a portion of a *Scaphite.*

Large quantities of ragstone from the upper greensand are quarried for building purposes around Selborne, and in these quarries large *Ammonites* are not infrequent.

BETWEEN NORE HILL AND THE HANGER HILL.

## CHAPTER IV

## "THE WAKES" AND ITS ASSOCIATIONS. SOME RELICS OF THE NATURALIST

GILBERT WHITE'S house at Selborne is known as "The Wakes." In a receipt for quit-rents made out in White's handwriting in 1787, the house is called "late Wakes," the previous owner or tenant having been known by that name.

The house in which the father of British popular naturalists lived and died, is situated in the village street of Selborne, in Hampshire, a few yards south of the "Plestor," or playground, which White has described in one of his early letters. As one enters the village from the direction of Alton, one crosses a little stream, which meandering down the vale called the Lithe, joins the stream which crosses the other end of the village, known as the Bourne. The latter at its source is familiar to all lovers of Gilbert White as the "Well-Head," and from this the water-supplies of the village are drawn. The church and vicarage stand on rising ground above the first-mentioned stream, and on the left-hand side of the road. They both face the Plestor, and are close to that corner of the churchyard where is found the famous yew-tree. "The Wakes," Gilbert White's house, stands on the opposite side of the road, a few paces beyond the Plestor. The wall of the narrow front garden, flush with the village street, was not there in White's time, and the space was occupied by shrubs.

"The Wakes" as it appeared when White lived there, seems to have been a picturesque house of the low-ceilinged, long and straggling type. From the numerous gables, and in consequence of the projection at the south-western corner of a suite of rooms beyond the general line of the back of the house, it appears to have been an eminently cosy and comfortable residence, sheltered at the rear by Selborne Down from the south-west Channel gales.

From the representations of the house which have adorned various editions of his book, it is not difficult to watch the changes which have taken place, gradual though they have been, in the fabric of the building. The alterations which have taken place are undoubtedly considerable, although on the other hand, as the expansion has been all in one direction, viz., at the northern end of the building, much of the old part does actually remain unaltered.

An illustration in Mitford's edition of 1813, and Jackson's woodcut in Professor Rennie's edition of 1833, show the house from the rear as it stood at White's death. This cut is reproduced in Harting's edition of 1891. In this picture there is a tall chimney, reaching to the ground, at the north termination of the house, and it is noteworthy that there are no windows of the type of what are now known as bay-windows. The chimney was no doubt built at the same time as the library, this being the room which White himself added, and which gave him so much concern in its building owing to the changes in the weather. In a letter to a sister-in-law at Blackburn, Mrs. John White, he wrote on April 7th (1778?):—

"My new parlor now dries at a great rate; and will be fit for use at Midsumr, but I shall not be able to compleat it this summer. I must not put on my upper paper 'til another year."

The next stage would appear to be that shown in Jardine's edition of 1853, where a wing has been added beyond the tall chimney, which now appears only as of ordinary length. The pathway leading to the summer-house is in this engraving roughly delineated. This path we shall deal with later on. In the "Introductory Observations," Sir W. Jardine refers to the fact that "one wing has been added since his death," and this is said to be in exact keeping with the other portions, with the exception, be it noted, that the new wing has bay-windows. This addition was made by Professor Bell.

Would that the "adding" process had stopped here. Photographs recently taken show further extensive additions. That which was formerly the final wing at the northern end, and which was but a single story, has had a second floor added to it, whilst to this, also, has been added a further wing or extension with first and second floors. These additions are unfortunately such as to make the older portions appear mean and unpretentious, in comparision with the arrogant-looking structure which has grown up beside it. Vegetation will in the course of time hide its bare walls. It is, however, greatly to be pitied that such extensive alterations were necessary. Even the long sloping roof in the centre of the building, as seen from the park, has been cut up to allow of a gable-window being inserted.

Upon Gilbert's death the estate passed into the hands of his brother Benjamin, the publisher of his book, and was for many years tenanted by their sister.

Professor Bell, so well known as the writer of *British Quadrupeds, British Reptiles, British Crustacea,* etc., lived at "The Wakes" from 1844 until his death in 1880, and exhibited the greatest care in retaining as far as possible the antique appearance of the house, although he added the north wing as already stated.

It subsequently came into the occupation of General Parr, who left it in 1892 to reside at Greatham. A Mr. Reid resided in it until 1895, when it was sold to Mr. Paxton Parkin, and it has been since occupied by Sir Edwin Pears and Col. Bibby.

There is still in and about the house much that may remind us of the naturalist. The room in which he wrote most of his correspondence is still to be seen, as also the room which witnessed the close of his peaceful career. His kitchen, a long, low-roofed room on the ground floor at the south-western end of the building, is still there, although now used as a sitting-room.*

The old portion of the house, as it appears from the village street, has met with little change, the most noticeable being, perhaps, the substitution of the wooden fence by one of stucco and tiles. Unless the project is carried through that the building should be purchased by the nation when next vacant, or by public subscription, there can be but one result, namely, that sooner or later the building will be thoroughly rebuilt, perhaps the older portion completely demolished, and "The Wakes" as it was in the 17th century become but a name.

To those who know Selborne, it would be a cause of infinite regret to find that the house had been, or was

*Since this was written the property has passed through various hands, and further alterations have taken place. An effort made to purchase it on behalf of the nation in 1902-3 failed.

intended to be, rebuilt. Selborne without "The Wakes" would offer no rallying-point for pilgrims, who, indeed, travel to it from all parts of the world.

In a copy of the original edition, for sight of which I am indebted to Mr. R. Adkin, of Hodeslea, Eastbourne, there is bound up the "Particulars and Condition of sale" of the estate on which "The Wakes" is situated, the sale having taken place on July 28th, 1840, at the Auction Mart, London, by direction of the executors of Mrs. Rebecca Terry. The sale was held by Mr. George Robins, of Covent Garden, and also included "A small Italian Villa," and "A Thatched Cottage Ornée." The description of the property is interesting. It states that "The Selborne Estate is now associated with the History of England, and the interest excited by this sale will necessarily be materially increased in public estimation by a delightful reminiscence, that it was within the hallowed walls of the property under consideration that the Rev. Gilbert White, the celebrated natural historian, achieved all his fame, thus giving an éclat to the village of Selborne, that hath for many a long year, placed it high in favourable opinion... His first edition of *The Natural History and Antiquities of Selborne* is a work destined, from the great simplicity of its style, the calm benevolence of its spirit, and the close observation evinced in every page, to be the most popular of any publication that has followed or preceded it." There is a good deal more in a similar vein.

As Professor Bell entered into residence in 1844, it would seem that he must have been the purchaser. It also seems to have become in a very neglected condition, for we are told that "it is at present fearfully in want of beautifying and ornamenting," and no doubt some little time elapsed before it was made reasonably habitable again.

From the house runs a brick path, four bricks wide the whole of its distance, which led the naturalist in unpleasant weather to his summer-house in the "park." This summer-house was situated beneath a circle of existing well-grown trees, which stand out alone as a single clump in the centre of the meadow. The path is as straight as possible until it reaches near to the site of the summer-house, when it bends northward, and forms a hook-shaped semicircle; here it evidently used to lead to the place where the door formerly opened. The summer-house has long since disappeared, but the path still remains. The bricks were taken up and relaid with scrupulous care in Professor Bell's lifetime, and Frank Buckland in one of his writings speaks of the loving care which Bell bestowed on this and on all other tangible reminiscences of White.

The sundial remains in its entirety. Like all other timepieces of the same class, this one bears evident traces of the lapse of years. One after another the years have rolled on, until nearly two centuries of sunshine have laid their shadow upon the venerable stone. Still, however, it tells the same tale. The brass dial upon which we can look, is the same as that upon which the observing naturalist looked. Even in the fresh air of Selborne there are enough of impurities to corrode and eat into the stone of which the structure is built. It stands between five and six feet high, but in a recently published picture of it, it appears even taller. This is, however, due to the fact that, as viewed from the lower level of the meadow, it is raised upon the higher ground of the lawn, which itself is level with the top of a low wall which surrounds the garden. Two large sarsen-stones are noticeable in the meadow, close to the Hanger.

It will be convenient here to speak of the various "relics" of Gilbert White which have come to light.

During the time the house was occupied by Professor Bell, he collected around him all upon which he could lay his hands, which had any connection with the former tenant of his house. Unfortunately at the sale which followed Bell's death in 1880, all these valued relics were dispersed. Miss Helen Watney of Liss wrote in *Nature Notes* (September, 1893) thus—"I attended Mr. Bell's sale and purchased, amongst other things, a small round table which, it was said, had belonged to White. There was no written authority given with it, and I have always doubted its ever having been used by a very little man to write on, for it is too high to sit at comfortably. I have greater faith in the authenticity of a glass spoon, a sort of tea-caddy spoon, and two tumblers, or glass mugs, with handles."

In connection with the centenary of the death of Gilbert White, a number of most interesting articles were exhibited at the conversazione of the Selborne Society on May 10th, 1893. These were:—A malacca cane with the White crest; a rushlight holder; a watch-stand; a pair of green spectacles in their case; a magnifying glass; Gilbert White's MS. Garden Kalendar, May 1st to November 16th, 1759—a careful record of garden work, with notes on the weather throughout the year; letter from Gilbert White to his brother at Blackburn, on family affairs and natural history items, August 12th, 1775; MS. of Letter III to the Hon. Daines Barrington; Sermon preached by the Bishop of Peterborough, Friday, February 15th, 1750, with Gilbert White's autograph—an interesting record of early missionary work; MS. in his own handwriting of one of Gilbert White's sermons.

The MS. of the sermon showed that it was preached from the 1st Epistle of John, 4th chapter, and 20th verse. It is numbered "99," and according to the endorsement it was preached by him on thirty-eight occasions between April 3, 1748 and 1792 (nine times at Selborne, and fourteen times at Faringdon). It then passed into the hands of his nephew, the Rev. Glyd White, who made a few alterations in it, substituted a new conclusion, and preached it at Brightwell, Oxon, in 1813. With some slight further alterations it was preached twice in that year by the Rev. Samson White, at Maidford and at Cold Higham, Northampton. Finally it was preached at Bassage, Glos. in 1852, by the Rev. F. H. White with fresh alterations.

This sermon, which has thus so often proved useful, was printed in 1863 in the *Journal of Sacred Literature*.

Two interesting documents, relics of the naturalist, were presented to the Selborne Society in May, 1896. These were (i) a tailor's bill, which it fell to Gilbert's lot to settle, on behalf of his deceased grandfather, Thomas Holt, and (ii) one of his manuscript sermons. I am enabled to reprint below a copy of the former; the latter it will not be necessary to reproduce here in full.

The tailor's bill, the spelling of which is particularly phonetic, runs thus:—

"Mr Gilbard White, Bill 1746.

| | | | | |
|---|---|---|---|---|
| May 27, | for a frock makin | 0 | 2 | 0 |
| | cote botens | 0 | 0 | 9 |
| | brest botens | 0 | 0 | 1 |
| | thred and tape | 0 | 0 | 3 |
| | cannes | 0 | 0 | 2 |
| May 30, | for a wescot makin | 0 | 2 | 6 |
| | twist and buckrom | 0 | 0 | 8 |
| | thred and silk | 0 | 0 | 4 |
| | cannes and tape | 0 | 0 | 3 |
| | | 0 | 7 | 0 |

July 11th, 1746. Received then of Gilbert White one of the executors of Tho: Holt: Esqr the above Bill in full of all Demands

I say to me, Charles Sharp."

(The last four lines are in Gilbert White's handwriting, with the exception of Charles Sharp's signature. The back of the bill is endorsed "Sharp the Taylors Bill for Long Cloaths.")

The sermon is preached from the Gospel of St. John, 11th chapter, 33rd and 35th verses. The essence of the sermon is contained in the first few sentences, and these are here reproduced.

"Many useful heads," said the preacher, "might be produced on this subject: but I shall content myself at present with the following: that grief, and pity, and a fellow-feeling of the miseries of others, and tenderness of heart, and bowels of mercy and love, and the rest of the human affections, are not inconsistent with wisdom and goodness: and that men *may* be truly wise, and virtuous, without being free from passions."

The sermon is endorsed as having been preached twenty times by the author, the first occasion being at "Selborne; September 17: 1758." This would seem to show that, although White was not curate at Selborne between 1752 and 1784, yet until his appointment at Faringdon in 1762, he occasionally occupied the pulpit at Selborne after death had removed his father in 1758. After being preached twice at Selborne (1758 and 1759) it was delivered thirteen times at Faringdon, when it again returned to Selborne, to be given to a new generation in 1784, 1786, 1788, 1790 and 1792.

Other interesting relics were at one time in the possession of the Rev. Edmund Field, of Lancing College, near Shoreham in Sussex. Mr. Field possessed portraits of Gilbert White, the grandfather of the naturalist, and of John White, his father. The late Earl of Stamford reported in *Nature Notes* that he had seen a stool which Gilbert White used in his study, and on which doubtless he sat when he wrote the "Selborne." "Evidently the naturalist enjoyed having free play for his arms; he had therefore sawn off the back of a carved chair, and had so turned it into a stool."

He also stated that the original of the vignette of the Hermitage which appears in so many of the earlier editions of Selborne, was in the possession of the Rev. Edmund Field, but that the original water-colour sketch ($11\frac{1}{2}$ in. by 8 in.) of the Hermitage, with the inscription "S. H. Grimm fecit, 1777," from which was taken this somewhat reduced and distorted oval vignette, was in the possession of the Rev. F. Gilbert White, vicar of Leusden, South Devon.

Rice Oxley mentions in his book on the Oxford Colleges that there is a chair of Gilbert White's in the library of Oriel College, and that the Common Room possesses his writing-table.

The most important of all, however, that remains to remind us of Gilbert White, viz., the manuscript from which the *Natural History and Antiquities of Selborne* was printed, were for the first time in its history offered for auction at Sotheby's, on April 26th, 1895. (They were not the original letters.)

The existence of the MSS. was only known to a few immediately connected with the White family, and the lack of information concerning it, painfully showed itself in the notes to some of the earlier editions of "Selborne." At Gilbert's death in 1793, nearly four years after the labour of his life was published, they passed into the hands of his brother Benjamin, the London publisher of the first edition. From him they descended to his son Benjamin, who bequeathed them to the Rev. Glyd White, from whom they descended to Mr. A. Holt White.

Familiarity evidently bred contempt, since for some years it was said they lay uncared for in a box in an outhouse, where these priceless treasures might have met with irreparable injury. However, they seemed to have suffered but little from such treatment, being in exceedingly good condition. The ink was but little faded, but the most noticeable feature was the regularity of the handwriting of old Gilbert. There were no signs of rush or hurry in his manner of writing. Leisure he had in quiet Selborne to the full, and leisurely he went about his pursuits. The writing of some hundred and fifty letters did not disturb the even tenor of his way, and twenty years or more were to elapse between his first conception of writing a book upon natural history, and the first publication of his only book, the one which we all know so well.

The lot, as it was submitted for auction, included all the Letters to Thos. Pennant (44); the first 62 (out of 66) of those written to the Hon. Daines Barrington; 12 pages of the "Antiquities;" 1 leaf of Index; 1 leaf of "More rare Plants Observed in the Parish of Selborne" (published in Bell's Edition); A Poem entitled "Selborne-Hanger, a Winter Piece, to the Miss Batties" (1 leaf); and another, "The Invitation: To Samuel Barker by the Rev. G. White" (2 leaves in a different hand); "On the Sense of Hearing in Fishes;" 2 leaves headed "Hints," but not finished (published only in Bell's Edition); and a leaf entitled "more particulars of the old family Tortoise." The manuscripts contained many passages not printed in the several editions, and had never previously been out of the possession of the lineal descendants of the author's brother.

A few sheets of the MSS. were written by an amanuensis, one of the author's family, but most of them were copies made by the author himself from his original letters (see page 31).

At the sale of the manuscripts there were few present, but much excitement was manifested amongst the twenty or twenty-five people in the sale-room. To the great astonishment of those present the first bid for the lot was no less than 200 guineas. This rapidly rose to 205 guineas and 210 guineas, when there was a pause. Bidding, however, commenced afresh, and rose by successive bids of five guineas each to 280 guineas, at which price the lot was knocked down to Messrs Pearson & Co., of Pall Mall Place, buying on commission. The MS. of the book was again auctioned in 1907, when it was sold for £750.

Seventeen autograph sermons of Gilbert White were then sold, and realised from eighteen shillings to thirty-six shillings apiece.

For reference purposes, some details of the sermons, extracted from Messrs Sotheby's catalogue, together with the prices at which many of them were sold, may prove useful:—

White (Rev. Gilbert): Autograph Sermon for Septuagesima Sunday, written in 1747, preached at Selborne and other Places from that date to 1778, sold for 26*s.*

Another for Epiphany, preached from 1748 to 1793, 25*s.*

Another for Easter, preached from 1749 to 1778.

Another preached from 1749 to 1773, 26*s.*

Another on Behaviour at Church, preached between 1749 and 1792, 25*s.*

Another for Christmas, preached between 1751 and 1790, 27*s.*

Another for Whit Sunday, preached between 1752 and 1793, 26*s.*

Another for Sacrament, preached between 1753 and 1791, 27*s.* (Maggs)

Another preached between 1754 and 1791, 21*s.* (Tregaskis)

Another preached between 1754 and 1791, 27*s.* (Maggs)

Another, preached between 1754 and 1792, 27*s.* (Edwards)

Another preached between 1757 and 1792, (Tregaskis)

Another preached between 1758 and 1792, 18*s.* (Maggs)

Another preached between 1758 and 1790, 36*s.* (Pearson)

Another preached between 1761 and 1791, 29*s.* (Pearson)

Another preached between 1762 and 1793, 22*s.* (Tregaskis)

Another preached between 1764 and 1789, 36*s.* (Maggs)

Another on Matthew 22, xiv (*no date*), 36*s.* (Edwards)

All were autograph sermons, and were endorsed as having been preached at Selborne, and in other places.

The sale was an extremely interesting one, and in view of the fact that no detailed account of the sale has appeared elsewhere, it would be as well to thus place on record an account by an eye-witness.

In the summer of 1895, the members of the Council of the Selborne Society had an opportunity of seeing other manuscripts, &c., of Gilbert White, at the residence of Mr. Rashleigh Holt White at Bexley. To some of these, reference has previously been made. They included:—

Thirty of the original letters to Mr. Pennant, 28 of which are signed. They bear on the back of each, the Alton postmark in large plain capital letters, thus ALTON, without any date. They have since been purchased for the British Museum, having been secured by Mr. Quaritch on that behalf at Sotheby's sale on Nov.

25, 1897, for £134. In the Museum Catalogue of MSS. they are numbered 35138. At the same time the Museum became the purchaser of White's "Garden Kalendar" in MS., dated from 1751 to 1767, at a cost of £177. It is numbered 35139.

White's College Account Book, in which appear such items as "Two botts. Rum, 5*s.*; Sugar and Lemmons, 6*d.*; Cutting Hair, 1*s.*; Loss at Cards, 2*s.*"

Account Books of his Selborne life, showing amongst many other things, payments which he made for the education of some poor children.

Letters from Rev. John White to Linnaeus, and copies of Linnaeus' replies.

Letter regarding a wager which White had with one of the Misses Batty (or as he terms them, the Miss Batties).

The Introduction to the missing *Natural History of Gibraltar* by his brother, the Rev. John White, consisting of 24 foolscap pages of MS. This was published by the Selborne Society in 1913 under the editorship of W. H. Mullens, M.A.

Gilbert's License to deacon's orders; to priest's orders; his Declaration on his induction into the living of Moreton Pinkney.

His "Garden Kalendar."

"Flora Selborniensis," containing also references to the fauna, for 1766. This evidently formed the basis of the "Naturalist's Calendar."

His barometer. His stick. Two seals.

*The Gentleman's Magazine*, containing the review of his book by his brother, Thomas White, a botanist of repute.

In April 1904 seventy-nine letters from Gilbert White to Mary White, and twenty-seven to his brother, the clergyman at Blackburn, were sold for £291. This excellent price brought on the market in June nineteen other letters to the same brother, and these realised £101.

A large collection of these and other relics of the naturalist was brought together at Mr. Holt White's later residence at St. Leonards, and were seen by me in 1927. After his decease, and that of his widow, a large proportion of them was dispersed at Sotheby's auction rooms.

A chapter dealing with relics of the naturalist would be incomplete which gave no particulars respecting the old family tortoise.

White deals with this in three of his printed Letters, viz: Nos. VII, XIII, and L to Daines Barrington, and he added further information under the title of "More particulars respecting the old family Tortoise," the last-mentioned being placed as an addendum at the end of the original edition of his book. Many short notices and references to the tortoise occur in various parts of his Diaries. The history of the creature as given by him is so interesting as to tempt a reprint of the letters and other references which deal with it.

RINGMER, *October* 8th, 1770.

"A land tortoise, which has been kept for thirty years in a little walled court belonging to the house where I now am visiting,* retires under the ground about the middle of November, and comes forth again about the middle of April. When it first appears in the spring it

*"Delves House," the house of his aunt Rebecca (born 1694); she married Henry Snooke, of Ringmer.

discovers very little inclination towards food; but in the height of summer grows voracious; and then as the summer declines its appetite declines; so that for the last six weeks in autumn it hardly eats at all. Milky plants, such as lettuces, dandelions, sowthistles, are its favourite dish. In a neighbouring village one was kept till, by tradition, it was supposed to be an hundred years old. An instance of vast longevity in such a poor reptile!"

Note in Diary:—

CHILGROVE, *November 15th*, 1771.

"Tortoise at Ringmer had not finished his hybernaculum, being interrupted by the sunny weather, which tempted him out."

*April 12th*, 1772.

"While I was in Sussex last autumn my residence was at the village near Lewes, from whence I had formerly the pleasure of writing to you. On the first of November I remarked that the old tortoise, formerly mentioned, began first to dig the ground in order to the forming its hybernaculum, which it had fixed on just beside a great tuft of hepaticas. It scrapes out the ground with its fore feet, and throws it up over its back with its hind; but the motion of its legs is ridiculously slow, little exceeding the hour-hand of a clock; and suitable to the composure of an animal said to be a whole month in performing one feat of copulation. Nothing can be more assiduous than this creature night and day in scooping the earth, and forcing its great body into the cavity; but, as the noons of that season proved unusually warm and sunny, it was continually interrupted, and called forth by the heat in the middle of the day; and though I continued there till the thirteenth of November, yet the work remained unfinished. Harsher weather, and frosty mornings, would have quickened its operations. No part of its behaviour ever struck me more than the extreme timidity it always expresses with regard to rain; for though it has a shell that would secure it against the wheel of a loaded cart, yet does it discover as much solicitude about rain as a lady dressed in all her best attire, shuffling away on the first sprinklings, and running its head up in a corner. If attended to, it becomes an excellent weather-glass; for as sure as it walks elate, and as it were on tiptoe, feeding with great earnestness in a morning, so sure will it rain before night. It is totally a diurnal animal, and never pretends to stir after it becomes dark. The tortoise, like other reptiles, has an arbitrary stomach as well as lungs; and can refrain from eating as well as breathing for a great part of the year. When first awakened it eats nothing; nor again in the autumn before it retires; through the height of the summer it feeds voraciously, devouring all the food that comes in its way. I was much taken with its sagacity in discerning those that do it kind offices; for, as soon as the good old lady comes in sight who has waited on it for more than thirty years it hobbles towards its benefactress with awkward alacrity; but remains inattentive to strangers. Thus not only 'the ox knoweth his owner, and the ass his master's crib,' but the most abject reptile and torpid of beings distinguishes the hand that feeds it, and is touched with the feelings of gratitude!

I am, &c., &c.

"P.S. In about three days after I left Sussex the tortoise retired into the ground under the hepatica."

"DELVES HOUSE," RINGMER, NEAR LEWES,
FROM THE REAR

RINGMER, *August 7th*, 1775.

"Timothy, Mrs Snooke's old tortoise, which has been kept full 30 years, in her Court before the house, weighs six pounds, three quarters, and one ounce. It was never weighed before, but seems to be much grown since it came."

CHILGROVE, *November 30th*, 1776.

"Mrs Snooke's old tortoise, at Ringmer, went underground."

*May 9th.*

"Timothy eats dandelion leaves and stalks. He swallows his food almost whole."

*June 4th.*

"Timothy took his usual ramble, and could not be confined within the limit of the garden. His pursuits, which seem to be of the amorous kind, transport him beyond the bounds of his usual gravity at this season. He was missing for some days, and found at last near the upper malt-house."

*June 20th.*

"We put Timothy in a tub of water, and found that he sunk gradually, and walked on the bottom of the tub. He seemed quite out of his element, and was much dismayed. This species seems not at all amphibious."

*March* 31, 1780.

"Timothy the tortoise accompany'd me from Ringmer. A jumble of 81 miles awakened him so thoro'ly that the morning I turned him into the garden, he walked twice the whole length of it to take a survey of the new premises, but in the evening he retired under the mould, and is lost since in the most profound slumbers."

SELBORNE, *April 21st*, 1780.

"The old Sussex tortoise, that I have mentioned to you so often, is become my property. I dug it out of its winter dormitory in March last, when it was enough awakened to express its resentments by hissing; and, packing it in a box with earth, carried it eighty miles in post-chaises. The rattle and hurry of the journey so perfectly roused it that, when I turned it out on a border, it walked twice down to the bottom of my garden; however, in the evening, the weather being cold, it buried itself in the loose mould, and continues still concealed.

"As it will be under my eye, I shall now have an opportunity of enlarging my observations on its mode of life, and propensities; and perceive already that, towards the time of coming forth, it opens a breathing place in the ground near its head, requiring, I conclude, a freer respiration as it becomes more alive. This creature not only goes under the earth from the middle of November to the middle of April, but sleeps great part of the summer: for it goes to bed in the longest days at four in the afternoon, and often does not stir in the morning till late. Besides, it retires to rest for every shower, and does not move at all in wet days.

"When one reflects on the state of this strange being, it is a matter of wonder to find that Providence should bestow such a profusion of days, such a seeming waste of longevity, on a reptile that appears to relish it so little as to squander more than two-thirds of its existence in a joyless stupor, and be lost to all sensation for months together in the profoundest of slumbers.

"While I was writing this letter, a moist and warm

afternoon, with the thermometer at 50°, brought forth troops of shell-snails; and, at the same juncture, the tortoise heaved up the mould and put out his head; and the next morning came forth, as it were raised from the dead, and walked about till four in the afternoon. This was a curious coincidence! a very amusing occurrence! to see such a similarity of feelings between the two φερέοικοι! for so the Greeks called both the shell-snail and the tortoise."

SELBORNE, *September,* 1787.

"Timothy, the tortoise, who has spent the last two months amidst the umbrageous forests of the asparagus beds, begins to be sensible of the chilly autumnal evenings."

In Bennett's edition of 1837 we read that the death of the tortoise at last occurred in 1794. Its carapace was preserved, and in 1836 was in the possession of Mrs. White. A woodcut of it illustrated the text of that edition. The creature lived at Ringmer for forty years, and at Selborne for fourteen years.

Bennett was extremely anxious to be able to announce that he had discovered certain distinctions which would justify him in investing the animal with a separate specific name, the name he proposed being "Whitei," after the naturalist. Professor Bell was, however, unable to recognise such distinctions, and announced that the species was undoubtedly *testudo marginata,* a species not uncommon in Greece and the Mediterranean.

From the correspondence which passed between the Rev. John White and Linnaeus, we learn that Gilbert was the first to detect the differences between the box-tortoises and those in which the sternum is not moveable upon a hinge. This, according to Bell, gave Linnaeus his first idea of the two distinct forms of the *testudinata*, so that although his brother John, in the correspondence referred to, does not appear to give Gilbert the credit of the discovery, yet to him it is undoubtedly attributable.

In speaking about the tortoise we must again refer to the amusing reply which White wrote to Miss Mulso, a daughter of his old college friend, John Mulso, concerning the journey which the tortoise was compelled to undergo, in travelling from Ringmer to Selborne. This letter, which appears to emanate from the pen of "Timothy, the tortoise, to Miss Hecky Mulso, from the border under the fruit wall," was first printed in Blyth's edition of 1850, and although subsequently appearing in several editions, does not seem to have received that recognition of its humour which it certainly deserves. This was apparently the letter which was sold in 1902 for £26.

The carapace of this once famous tortoise has been honoured by a place in one of the cases of the National History Museum at South Kensington. It was presented by Mrs. Christopher, White's great-niece, to the British Museum in 1853. One or two of the scales are missing, otherwise it appears to be a fair representation of the tortoise one frequently sees hawked about the London streets. The specimen bears the inscription: "The Shell of Gilbert White's Tortoise, *Testudo Ibera*," and we are informed in the label attached to it that it died in 1794, having lived about 54 years in England, the last 14 of which were spent at Selborne. It is conveniently exhibited in the reptilian and amphibian room at the Museum.

STONE COFFIN FROM SELBORNE PRIORY.

# CHAPTER V

## THE TESTIMONY OF MEN

THE testimony which has been borne to Gilbert White's *Natural History and Antiquities of Selborne* by writers distinguished in all walks of literary life, is to be found on all hands scattered throughout the muniments and archives of the country.

There can be no doubt that the book has exercised considerable influence on leading men in all generations. The late American statesman and eminent literary man, James Russell Lowell, in his "My Garden Acquaintance," the first essay in *My Study Windows*, wrote as follows:—

"One of the most delightful books in my father's library was White's *Natural History of Selborne*. For me it has rather gained in charm with years. I used to read it without knowing the great pleasure I found in it; but as I grow older I begin to detect some of the simple expedients of this natural magic. Open the book where you will it takes you out of doors. In simplicity of taste and natural refinement it reminds one of Walton, and in tenderness of Cowper . . . The book has also the delightfulness of absolute leisure. Mr. White seems never to have had any harder work to do than to study the habits of his feathered fellow creatures, and watch his peaches ripening; his volumes are a journal of Adam in Paradise. It is positive rest only to look at that garden. There are moods in which this kind of history is infinitely refreshing. These creatures whom we affect to look down upon as the drudges of instinct, are members of a commonwealth whose constitution rests on immovable bases."

Again, in another place, "it is good for us now and then to converse with a world like Mr. White's (of Selborne) when man is the least important of animals.... It is an innocent and healthful employment of the mind distracting one from too continual study of himself, and leading him to dwell rather upon the indigestion of the elements than his own."

This is indeed valuable testimony, and well describes the growth of the appreciation which has been experienced by many of White's admirers. The value of the book is not, it always seems to me, apparent at first reading, much less, perhaps, when read in boyhood's or girlhood's days. The growth of its appreciation increases rapidly with increasing years.

On a visit paid by Mr. Lowell to Selborne in 1880, in the company of Lady Sophia Palmer, daughter of the late Lord Selborne, the poet composed some lines appropriate to the occasion, which his lordship read before the Selborne Society. They were published in *Nature Notes*, in October, 1893, and are as follows:—

To visit Selborne had been sweet
   No matter what the rest might be;
But some good genius led my feet
   Thither in such fit company,
As trebled all its charms for me.

With them to seek his headstone grey,
   The lover true to birds and trees,
Added strange sunshine to the day.
   My eye a scene familiar sees,
And Home! is whispered by the breeze.

My English blood its right reclaims;
In vain the sea its barrier rears;
Our pride is fed by England's fame,
Ours is her glorious length of years;
Ours, too, her triumphs and her tears.

It is extremely gratifying to know that in America the name of Gilbert White is held in the same honour as in our own country.

The following is a copy of a letter from the late Duke of Argyll.

"In a small volume of poems I published last year (John Murray, Albermarle Street), there is one short poem on 'Selborne' expressing in detail the effect that Gilbert White's book has always had on me from my earliest years. That impression corresponded closely with the description given by Lowell in the extract you send to me. I had not seen it before. The Rest and Peace of Nature is the Charm. My book is entitled: *The Burden of Belief, and other Poems.*"

Professor Bryce, M.P., wrote: "I join heartily in the admiration for White's 'Selborne.'"

The following important letter is from Sir Edwin Arnold, author of *The Light of Asia*, *The Light of the World*, etc.

"I grieve that I have now no time, or sufficient space to express the literary and intellectual debt which I owe to dear old Gilbert White of Selborne. His was certainly one of the books from which I learned my own love of Nature and her children, and taught the same to my own boys. To criticise is impossible, where affection and gratitude are the emotions awakened. Above all, he shewed me the beautifulness of little and humble things."

This is, indeed, a sweet letter of testimony to greatness from greatness, and is worthy of permanent record.

Surely, too, the following poem, which was published in *The Speaker* of June 17th, 1893, deserves, as a tribute to our hero, a place in such a work as this:—

IN THE COUNTRY OF GILBERT WHITE

GHOSTS of great men in London town
  Confuse the brains of such as dream,
But here betwixt this hanging down
And this great moorland, waste and brown,
  One only reigns supreme.

In Wolmer Forest, old and wide,
  Along each sandy pine-girt glade
And lonesome heather-bordered ride,
A gentle presence haunts your side,
  A gracious reverend shade.

And as you pass by Blackmoor grim
  And stand at gaze on Temple height,
Methinks the fancy grows less dim:
Methinks you really talk with him
  Who once was Gilbert White.

For yonder lies his own true love,
  His little Selborne, dreaming still:
The shapely "Hanger" towers above,
Girt with its beautiful beech grove
  Like some old Grecian hill.

And there th' abrupt and comely "Nore"
  Guards that wild world of bloom and bird
Where his clear patient sense of yore
Conned sights and sounds, which ne'er before
  Sweet poets saw or heard.

And here, hard by, the nightingale
  For the first time in springtide sang,
While Gilbert listened; here the pale
First blackthorn flowered, while down the gale
  The cuckoo's mockeries rang!

And there rathe swallows would appear,
  To whirl on high their first gavotte;
And there the last of the great deer
Fell on a winter midnight clear.
  'Neath a "night-hunter's" shot.

We know it all! Familiar, too,
  Seems this quaint hamlet 'neath the steeps—
House, "Pleystor," church, and churchyard yew,
And the plain headstone, hid from view,
  Where their historian sleeps.

'Twas just a century gone by
  They laid the simple cleric here:
Th' old world was in her agony,
And "Nature! Reason!" was the cry
  In that historic year.

But O! another Nature 'twas
  That ruled him with her magic touch,
A mistress of delightful laws,
Whom still we learn to love because
  We love her servant much!

VICTOR PLARR

In Miss Mitford's well-known book *Our Village* she has a highly appreciative reference to Gilbert White. "Nothing," she says, "is so delightful as to .... ramble with Mr. White, whose *Natural History and Antiquities of Selbourne* is one of the most fascinating books ever written, over his own parish, and form a friendship with the fields and coppices, as well as with the birds, mice, and squirrels who inhabit them."

It is probable that the influences of White's life and work have had considerable effect on all the leaders in natural history science during the present century. Richard Jefferies in his preface to a comparatively recent edition of "Selborne," abundantly testifies to this. He compares White to the great painter Turner, who saw in the majestic downs the beauties which White bore witness to, and about which he wrote:—

"For my own part, I think there is something peculiarly sweet and amusing in the shapely figured aspect of chalk hills in preference to those of stone, which are rugged, broken, abrupt, and shapeless. Though I have now travelled the Sussex Downs upwards of thirty years, yet I still investigate that chain of majestic mountains with fresh admiration year by year; and I think I see new beauties every time I traverse it."

Richard Jefferies goes on to say, "He thinks them glorious, as indeed they are: yet one would scarcely expect, in the world's judgment, a man who was not an authority on art, to find out for himself the views which the public now purchase so eagerly. The sympathy he felt with nature enabled him to see much further than the hopes by which he walked, and brought his mind into parallel lines with the great painter."

We may fairly be thankful that White never had to pass examinations for scientific degrees before he was able to make his mark on the natural history world. Otherwise the poetry and the sublime art of nature must have been crushed out of him, in the useless striving after distinction at the examination table. Naturalists are born indeed; they cannot be made.

In view of the comparison which has been made between Darwin's work on the place of worms in creation, and the letter of White's dealing with the same subject a hundred years before, the remarks in the *Life and Letters of Charles Darwin* bearing on Gilbert White's life and influence are extremely valuable.

Darwin wrote that when about sixteen years old, "From reading White's Selborne, I took much pleasure in watching the habits of birds, and even made notes on the subject. In my simplicity, I remember I wondered why every gentleman did not become an ornithologist." As it has been remarked, it appears very probable that this making of notes on the subject, was really the first of the great scientist's writings, and that his attention was drawn in that direction in the first place by the reading of Gilbert White's book. Darwin is stated by his son to have made a 'pilgrimage to the shrine of Gilbert White, of Selborne,' between 1856 and 1858.

Mr. Edward Jesse, who edited White's book in 1851, had the naturalist's diaries in his possession for some time, and made the following appreciative remarks in his *Gleanings in Natural History*. He says, "White notices the visits which were from time to time paid him, and these were chiefly by his own relations. He carefully notes down all his nephews and nieces as they respectively came into the world, amounting to about 63 when his diary closed. He chronicled his beer, and takes notice of the quantity of port wine which came to his share when he had a pipe of it with his neighbours."

"He appears to have made annual visits to Lambeth, Ringmer, and Oxford, and during these absences, his old servant, Thomas, kept his journal of the weather. Mr. White passed much of his time in his garden, and he makes frequent mention of his crops, his fine and early cucumbers, and the flavour of his cardilliac pears. He thought nothing unworthy of being recorded which was connected with rural life. The appearance of his neighbours' hops, the beginning and ending of their harvests, their bees, their pigs and poultry, are all noticed in successsion, and appear to have added to the interest he took in rural life. Insignificant as these little details may appear, they were not thought so by a man of cultivated and elegant taste, whose mind was stored with learning, and who has left behind him one of the most delightful books in the English language."

In Jesse's edition (1851) the following poem by his daughter appears:—

He lived in solitude—'midst trees and flowers
Life's sunshine mingling with its passing showers
No storms to startle, and few clouds to shade
The even path his christian virtues made.
Yet not alone he lived! Soft voices near,
With whispered sweetness, soothed the good man's ear;
He heard them murmuring through the distant trees,
While, softly wafted on the summer breeze,
The hum of insects and the song of birds
Spoken to his heart in tones more sweet than words.

Him in those quiet shades the poor might bless,
Though few intruded on his loneliness;
He fed the hungry, pitied the distress'd,
And smooth'd their path to everlasting rest.
Thus hearing Nature speak in every sound,
Goodness and love in all her works he found,
Sermons in stones and in the running brooks:
Wisdom far wiser than in printed books,
And in the silence of his calm abode
In nature's works he worshipp'd nature's God.

MATILDA HOUSTON

In the *Yale Review* for 1926, two letters were printed which were written by Thomas Carlyle to Eckermann, and I am indebted to the Provost of Oriel College for copies of them. They form part of the collection of Yale University Library.*

2 *August*, 1848.

"Do you know a little English book we have called White's Natural History of Selborne. It should be in the Duke's Library. If it is not accessible, tell me, and I will get it sent to you. You should do something like it for your own corner of Germany; you really should. It is one of our most excellent books; White, a quiet country Parson, has preached a better sermon here than all the Lord Bishops three times over. Think of it, I pray you."

5 *Jan.*, 1849.

"I have a reasonable little copy of White's Nat'l History of Selborne, which I will send to the Bookseller

*Reprinted from the *Yale Review*, July, 1926, copyright Yale University Press, by permission of the Editors.

Nutt this day, with a request to have it forwarded to you by the shortest route he has. I shall rejoice to hear that it arrives soon and safe; some sympathy in reading it I can anticipate from you; and if by example it kindles you into doing the like for Weimar and the environs, truly it will be a copy of a book well bestowed! White was a Peaceable Country Clergyman; his Parish Selborne is in Hampshire on the South Coast of this Island —in that same region where William Rufus, the "Red King," whom you may have heard of, was shot, hunting long ago. Of all the books of Natural History I ever read, White's, in his healthy genial limitation, pleases me most. Go and do likewise!"

# CHAPTER VI

## BIBLIOGRAPHY—FIRST EDITION, 1789

SO many editions have been published of the well-known *Natural History and Antiquities of Selborne* since the original 4to volume appeared in 1789, that it is interesting to take a survey of the various subsequent editions of the book which have been issued and to notice the treatment which it has received at the hands of its numerous editors and annotators.

Amongst them we may mention Markwick, Aikin, Mitford, Jardine, Rennie, Brown, Bennett, Wood, Harting, Buckland, Bell, Burroughs, Allen, Sharpe, Miall, Fowler, and Kearton.

Aikin and Markwick did not, strictly speaking, edit a new edition. The *Naturalist's Calendar*, the *Observations on Various Parts of Nature*, and the *Summary of the Weather*, were extracted from White's MSS. by Dr J. Aikin, and these were published as a separate book in 1795, six years after the original work was published, under the title of *A Naturalist's Calendar*. The material thus made public was added to the "Natural History" when a second edition was called for in 1802.

The original 4to volume was published in 1789, by "B. White & Son, at Horace's Head, Fleet Street." It is remarkable that the author's name does not appear on the title-page. We are informed in an advertisement at the end of the *Naturalist's Calendar* that the price of the first edition was one guinea. From the catalogue of Benjamin White I find that it was issued in three styles of binding at 21*s*., 24*s*., and 29*s*. A copy in the original boards was sold in 1900 at Sotheby's to Tregaskis for £7 2*s*. 6*d*. This was out of the library of the late Mr. C.

Leeson Prince, of Crowborough. He informed me that he gave fifteen shillings for it in 1864.

The book consisted of 468 pages, exclusive of the index. It contained seven copperplate illustrations, with a further illustration on page 307. Page 292 is numbered "262," and there are no pages 441 and 442. There are four copies in the British Museum.

A large folding engraving faces the engraved title-page, and is entitled "North East view of Selborne, "from the Short Lythe; Published Nov. 1, 1788, as the "Act directs, by B. White & Son." This large engraving is reproduced in the 4to edition of 1813, on a small scale in Jardine's edition of 1853, in the reprint of 1880, and in Mosley's edition of 1905. The four figures in the forefront of the picture were said by Professor Bell to be, commencing on the left hand, (i) Rev. Robert Yalden, (ii) Rev. Wm. Etty, (iii) Mrs. Yalden (afterwards Mrs. Benjamin White), (iv) Mr. Thomas Holt White, Gilbert White's brother.

The title-page runs thus:—"The Natural History "and Antiquities of Selborne, in the County of South-"ampton: with Engravings, and an Appendix.... 'ego "Apis Matinæ More modoque Grata carpentis ... per "laborem Plurimum ..... HOR. Omnia bene de-"scribere, quæ in hoc mundo, a Deo facta, aut Naturæ "creatæ viribus elaborata fuerunt, opus est non unius "hominis, nec unius aevi. Hinc *Faunæ & Floræ* utilissi-"mæ; hinc *Monographi* præstantissimi.' SCOPOLI ANN. "HIST. NAT. London: printed by T. Bensley; for B. White "and Son, at Horace's Head, Fleet Street, MDCCLXXXIX."

The back of the title-page is blank. Pages iii, iv, and v are devoted to Gilbert White's "Advertisement," which is signed "Gil. White, Selborne, January 1st, 1788."

Catch-words are used at the right hand bottom corner of each page, except when the page is complete in itself.

Page vi is blank. On page vii is the title to the first part of the book and runs thus:—"The Natural History of Selborne," beneath which is an engraving of the Hermitage after a drawing by S. H. Grimm; "D. Lerpinière sculp." The legend beneath the engraving is a quotation from White's own poem *The Invitation to Selborne*, and runs thus:—

"Where the Hermit hangs his straw-clad cell."

The poem, it should be mentioned, was at the time in manuscript only, not as yet having been published.

Beneath this are quotations from Homer and Cicero; back of page blank.

The Letters to Thomas Pennant, Esq., then follow, extending from pages 1 to 114.

Facing page 7 is inserted the engraving of "Mytilus, Crista Galli" in three positions: "published 1 Nov. 1788, as the Act directs, by B. White & Son." A folding engraving of "Charadrius, Himantopus" faces page 259.

The Letters to the Hon. Daines Barrington commence on page 115, and finish on page 305.

Title to "The Antiquities of Selborne" is found on page 307, with the Seal of the Priory in a border bearing the legend, "Beate Marie D'Seleborn Sigiil Conventus Eccle." Below the Seal is a quotation from Virgil, "Juvat ire .... desertosque videre locos."

The first letter of the "Antiquities" commences on page 309, and the last concludes on page 427.

An engraving of the "South View of Selborne Church." "P. Mazell Sculpt., S. H. Grimm delt.," with the old yew tree in the foreground, appears opposite to page 315.

An engraving of the "North View of Selborne Church," also by Grimm and Mazell, with the Vicarage on the right, appears opposite to page 323.

An engraving of "Temple, in the Parish of Selborne," appears opposite to page 343.

An engraving of "The Pleystow, vulg: the Plestor," appears opposite page 345.

On pages 427 and 428 are found "More particulars respecting the old family tortoise, omitted in the Natural History."

On pages 429 and 430 are found "Additions to Letter XLI, page 236." This is the paragraph commencing, "Of all the propensities of plants," and in all subsequent editions it has been placed at the end of the letter referred to. It concludes with the poetic lines commencing, "Say, what impels, amidst surrounding snow Congealed, the *crocus*' flamy bud to glow?"

The appendix is found on pages 431-468, and consists of copies of Latin documents relating to the Priory and neighbourhood.

Index follows on pages 469-480, although these few pages are not numbered.

A "List of Plates to the History of Selborne" appears on page 480. This reference and that on the next page, appear to be the only two made to the whole book as a "History of Selborne."

On the next page, the last printed page, is a list of

"Errata in the History of Selborne," consisting of 18 corrections.

The first 15 of these errata are referable to the Natural History, and the necessary corrections have been made in the succeeding edition of 1802, with the exception of "Ward-le-ham," (Letter XLV to Barrington) which is in that edition still spelt "Wordleham." The three remaining errata refer to the Antiquities. This part of the book was not reprinted with the 1802 edition. On referring to the edition of 1813, it will be found that the erratum at the end of Letter X has been incorrectly treated. In the original edition we read in the Errata, "Page 352, line 12, for 'esteemed' read 'deemed.' " In that edition the word 'esteemed' occurs both on line 11 and line 12, but the compilers of the 1813 edition have altered the word on line 11, and have left it untouched on line 12, whereas it should have been untouched on line 11, and altered on line 12.

This inattention to the errata shows itself in every subsequent reprint of the "Antiquities." Rennie's edition is at fault in this respect, and so is Blyth's and Jardine's, and Harting's and Bell's.

Excluding the index and the preface, which is called in this and other old volumes the "Advertisement," there are 468 pages (less 2, namely pages 441 and 442). It is worthy of note, in view of the arrangement in some subsequent editions, that the forty-four letters to Pennant, and the sixty-five letters to Barrington, follow each other in two separate series, although the dates of the letters in the one run contemporaneously for some years with those in the other. The first dated letter to Pennant (No. X) bears date August 4th, 1767, and the last November 30th, 1780; whilst the first and the last to Barrington are dated June 30th, 1769, and June 25th, 1787, respectively.

After the last letter of the "Antiquities" there are two additions to which I wish to call special attention. They appear to have been overlooked by some editors as having actually been printed in the first edition. The first is "More particulars respecting the old family tortoise, omitted in the Natural History." This, in the second edition of 1802, and in many of the subsequent editions, has been placed at the end of Letter L to Barrington, in which other mention is made of the tortoise.

The other addition is headed on page 429 "Additions to Letter XLI, page 236," and commences "Of all the propensities of plants." The paragraph refers to the times of flowering of the vernal and autumnal crocuses, and closes with a few lines of poetry. It is remarkable that this addition appears to have been totally overlooked by subsequent editors, as occurring in the first edition. Sir William Jardine states in his 1853 edition that it was added in Mitford's edition of 1813, "but we are not aware at what time or under what circumstances this was written." Even Professor Bell in his valuable edition of 1877, says that it was first printed in the edition of 1802. Then again, Grant Allen wrote in 1899, "I do not know whence Mitford derived it." The fact of the matter is that it did appear in the first edition, in 1789, but it was printed out of place, viz., at the end of the "Antiquities."

In looking through some old book-catalogues at the British Museum Library, I found five catalogues which were issued by the brothers of Gilbert White, viz.,

Benjamin and John White. A transcript of the title-page and other particulars are of interest. That dated 1795, runs as follows: "A Catalogue of an extensive "and valuable collection of books; containing many "recent purchases of rare and valuable articles." "The "sale will begin on Monday, the 12th of January, 1795, "by Benjamin and John White, London.—N.B. The "lowest prices are marked in the catalogue, and on the "first leaf of every book."

The catalogues of those days were substantial, there being in this case no less than 10,496 items, contained on 332 pages.

On page 122, the following appeared: "3,361: "*White's Natural History, &c., of Selborne, in the County* "*of Hants, in Letters to Mr Pennant, and to Mr Barrington,* "fine plates, boards, £1 1*s*. 3,362: The same, new and "neatly bound, £1 4*s*. 3,363: The same, elegantly bound "by Kalthoeber, £1 9*s*."

There were four other similar catalogues.

In 1791 a catalogue of 10,658 items were offered for sale on 25th January, and on page 107 the same three books appeared, numbered 3,029, 3,030, and 3,031. In 1792 the date of the sale was February 13th, and in a list of 10,940 books, the three White books appeared on page 111, numbered 3,087, 3,088, and 3,089. In 1793 the same were again for sale on February 11, and in a list of 10,334 books, those of White appeared on page 114, numbered 3,105, 3,106, and 3,107. Then again in 1796, a sale of 10,809 items took place on January 18, and the three books constituted numbers 3,177, 3,178, and 3,179, on page 116.

It will thus be seen that Gilbert White's book was

issued in three styles of binding, and that it regularly appeared in the publishers' catalogue some years after the naturalist's death.

Second-hand Catalogue Prices: £3 17*s.* 6*d.*, £9 9*s.*, £17 10*s.*, £7 2*s.* 6*d.*, £17 10*s.* (uncut), £18 (1926).

From *Book Auction Sales* and *Book Prices Current* (by permission), and other sources:—

Auction Prices: 1887, £1 8*s.*, £2 8*s.*: 1888, £2: 1889, £1 18*s.*: 1890, £2 12*s.*: 1891, £2 12*s.*: 1893, £10: 1894, £10 15*s.*: 1896, £4, £5 12*s.* 6*d.*: 1898, £6 10*s.*: 1899, £8 8*s.*, £19 15*s.* (uncut): 1900, £7 2*s.* 6*d.* to £17 10*s.*: 1901, £6 7*s.* 6*d.* to £11: 1902, £9 12*s.* 6*d.* to £30: 1903, £11 2*s.* 6*d.* to £31: 1904, £8 10*s.* to £31, and £22 (large paper, one of 50): 1905, £9 5*s.*, £20: 1906, £8 8*s.*, £26 10*s.*: 1907, £9, £11 5*s.*: 1908, £7 10*s.*, £13 10*s.*: 1909, £5 to £15 15*s.*: 1910, £6 15*s.* to £20: 1911, £1 11*s.* (rebacked), £4 to £10 10*s.*: 912, £7 10*s.*: 1913, £10 to £18: 1914, £3 15*s.* to £8 8*s.*: 1915, £4 2*s.* 6*d.* to £25: 1916, £5 5*s.*, £14: 1917, £7 15*s.*, £10: 1918, £7 5*s.*, £18 10*s.*: 1919, £8 10*s.* to £25 4*s.*: 1920, £8, £10 10*s.*: 1919, £8 10*s.* to £25 4*s.*: 1920, £8, £10 10*s.*, £21: 1921, £8 to £12: 1922, £8 to £14 10*s.*; one fine copy £21; damaged copies, £5 5*s.*, and £3 15*s.*: 1923, £6, £8 5*s.*, £12 12*s.*, £23; damaged copy, £5: 1924, £12 10*s.*, £17; damaged, £5 10*s.*, and £7 5*s.*: 1925, £8 10*s.*, £11 10*s.*, £12, £13; damaged, £6 10*s.*, and £7 10*s.*: 1928, £24: 1929, £30: 1930, £14, £23: 1931, £43, (uncut): 1932, £6 5*s.*, £24: 1933, £6, £9.

# CHAPTER VII

## EDITIONS OF AIKIN AND MARKWICK, MITFORD, JARDINE (in part), AND RENNIE

GERMAN EDITION OF 1792. Published by F. A. A. Meyer, of Berlin. 16mo. 168 pages.

Title-page. "White's Beyträge zur Naturgeschichte "von England. Aus den Englischen übersetzt und mit "Anmerkungen begleitet von Friedrich Albrecht "Anton Meyer, . . . . . Berlin, 1792."

According to Newton, this edition was incomplete and inaccurate, consisting of extracts loosely put together, and "the editor's remarks often convey a sneer."

Auction Price: 1931, £13.

A NATURALIST'S CALENDAR was first published in 1795, as a separate volume, 8vo.; and is stated in the title page to have been "Printed for B. & J. White, Horace's Head, Fleet Street."

There is a short, but interesting "Advertisement" by Dr. Aikin on pages iii and iv, signed "J. Aikin, London, Jan. 1, 1795."

The volume includes:—

(i.) "The Naturalist's Calendar," and at the head of it are a few rematks which show the labour involved in compiling it from Gilbert White's journals. We find that in respect of the flowering of many plants only one entry appeared throughout the journals, this being in the year 1768, which, we are told, was a rather backward year. This is worthy of note.

(ii.) "Observations on Various Parts of Nature," which commences on page 55.

A coloured picture of "A Hybrid Bird" appears opposite to page 65. In small characters beneath appear, "Elmer pinxt.—London, Published April 14, 1795.—J. F. Miller, sculpt." Prof. Newton wrote of this illustration that "it was so badly done as to misre-"present not only the original but also the water-colour "drawing from which the plate is copied." I saw the original at Mr. Rashleigh Holt-White's residence at St Leonards in 1927, and can bear out fully Prof. Newton's remarks.

(iii.) "Meteorological Observations," which commence on page 140.

(iv.) The Poem, "On the dark, still, dry, warm weather occasionally happening in the winter months," which appears for the first time on pages 148-9.

(v.) "Summary of the Weather," which is found on pages 151—170.

A "Table of Contents" appears on the five following (unnumbered) pages.

On the back of the last page appears the following notice:—"Published by the same Author, In a Series of "Letters to the Honourable Daines Barrington, and "Thomas Pennant, Esq. In one Volume Quarto, price "One Guinea, in Boards; The Natural History and "Antiquities of Selborne, in the County of Southamp-"ton, With nine Engravings and an Appendix."

Note.—Long "s" is used in this volume, when that letter occurs anywhere but at the end of a word.

Second-hand: 7*s.* 6*d.*, 14*s.*, 8*s.* 6*d.*, 5*s.*, 22*s.*, 10*s.* 6*d.*, 24*s.*, 15*s.*

Auction Prices: 1923, 10*s.*, 15*s.*

A LARGE-PAPER EDITION OF A NATURALIST'S CALENDAR was published in quarto. The copy which I have examined bears the words "The Revd. Samson White, from John and James White."

Auction Prices: 1924, £1 12*s*.: 1925, £5 15*s*.: 1926, £2 4*s*.

EDITION OF 1802, ARRANGED BY DR J. AIKIN. Markwick's Calendar is added. The "Antiquities" are omitted. The old style of long "s" is used in this edition. In Two Volumes. Two coloured engravings.

*Volume I.* First page unnumbered. "The Works in Natural History," etc., etc.

The second page, according to the numbering, although no number appears upon it, contains an engraving of "Where the Hermit hangs his straw-clad cell." "S. H. Grimm, del." appears on the left, "D. Lerpinière, sculpt." on the right of the engraving, which is oval in outline. Beneath the oval are the words "Published as the Act directs, by J. White, Jany. 1st, 1802."

Page iii. Title-page. "The Works in Natural History "of the late Rev. Gilbert White, A.M., Fellow of Oriel "College, Oxford. Comprising The Natural History "of Selborne; The Naturalist's Calendar; and Miscel-"laneous Observations, *Extracted from his Papers*.

"To which are added, a Calendar and Observations, "by W. Markwick, Esq., F.L.S.;* in two Volumes. "Volume I. London: Printed for J. White, Fleet Street, "By T. Bensley, Bolt Court, 1802."

*Marwick's MS. Calendar is now deposited in the Public Library, Brassey Institute, Hastings.

Page iv remains blank.

Page v. Advertisement (Preface) by "J. W.," followed on page vii by a "few biographical records of the author," signed "J. W., Fleet Street, 1802." John White, the writer, was son of the elder Benjamin, and nephew of Gilbert.

On page 1 of the letters commences "Letter I to Thomas Pennant, Esq."

This edition has catch-words on the right-hand bottom corner of each page.

Between pages 12 and 13 are inserted three representations of the fossil curiosity, "Mytilus, Crista "Galli, published as the Act directs, by J. White, January 1st, 1802." In the right-hand bottom corner of the engraving are the words "Volume I, Page 12."

The last page of Volume I is numbered 392, there being at foot "T. Bensley, Printer, Bolt Court, Fleet Street, London."

*Volume II.* Opposite the title-page, is a folding, coloured engraving of "*Charadrius himantopus,* published as the Acts direct," etc.

Title-page same as in volume I.

Letters commence in this volume with Letter XLI, "To the Honourable Daines Barrington."

At the end of Letter L appears "More Particulars respecting the Old Family Tortoise" (on pages 49, 50 and 51).

The last letter of the "Natural History" ends on page 119.

"Antiquities" omitted in this edition.

On page 121 is a title-page to the second part of the book; "A Comparative View of the Naturalist's Calen-

"dar, as kept at Selborne, in Hampshire, by the late "Rev. Gilbert White, M.A., and at Catsfield, near "Battle, in Sussex, by William Markwick, Esq., F.L.S., "from the year 1768 to the year 1793."

On page 122 appears the note in the middle of the page: "N.B.—The dates in the following calendars, "when more than one, express the *earliest* and the *latest* "times in which the circumstance noted was observed."

Between page 122, and page 123, is inserted a coloured plate of "A Hybrid Bird."

The "Calendars" commence on page 123.

Page 157. "Observations on various parts of Nature, from Mr White's MSS., with remarks by Mr Markwick."

Page 158. Blank.

"Observations on Birds" commence on page 159. Remarks on "A Hybrid Pheasant" appear on page 172 and following pages.

On page 179, the word "turkeys" is spelt "turkies."

The poem "On the dark, still, dry, warm weather," etc., appears on pages 267-8.

The "Summary of the Weather," pages 269—285.

The index commences on page 287.

Total number of pages 300. At foot of last page "T. Bensley, Printer, Bolt Court, Fleet Street."

Samuel Taylor Coleridge's copy of this edition, with MS. notes, is in the British Museum. The notes are quoted in Grant Allen's edition of 1899.

In some copies the two volumes are bound together.

Second-hand: 15*s.*, 5*s.*, 13*s.* 8*s.*: 1932, in one vol, 15*s.*

Auction Price: 1927, £1 1*s.*

MITFORD'S 4TO EDITION OF 1813. One volume quarto, £2 12*s*. 6*d*. The new style of "s" has been adopted in this edition. First printed page, "The Natural History and Antiquities of Selborne, &c., &c." Back of same page, near the left-hand bottom corner, "T. Bensley, Printer, Bolt Court, Fleet Street."

A large folding plate is inserted here, entitled "North-east view of Selborne, from the Short Lythe." Title-page. "The Natural History and Antiquities of Sel- "borne, in the County of Southampton. To which are "added, The Naturalist's Calendar; Observations on "Various Parts of Nature; and Poems. By the late Rev. "Gilbert White, formerly Fellow of Oriel College, "Oxford. A new Edition, with Engravings. London: "Printed for White, Cochrane & Co.; Longman, "Hurst, Rees, Orme, and Brown; J. Mawman; S. Bag- "ster; J. & A. Arch; J. Hatchard; R. Baldwin; and T. "Hamilton, 1813." Back of title-page blank.

"Advertisement" on pages v, vi, vii. This is Gilbert White's original preface, dated "Selborne, January 1st, 1788."

Pages viii and ix. John White's preface, the same as in the second edition (1802). On page ix is a half-page engraving, "View of the Residence at Selborne, of the late Rev. Gilbert White. J. Harris, delt.; W. Angus, sculpt."

Page x. "Advertisement to the new Edition." This is also by "J. W.," and is dated "Selborne, May 10, 1813." Reference is made to the "elegant remarks" of the Rev. John Mitford, of Benhall, his notes first appearing in this edition. A note to this page states that Mr. Markwick, subsequently to the publication of his

Calendar, changed his name to Eversfield, on succeeding to his uncle's estate, and died at Catsfield, on the 6th of April, 1813.

Page xi contains an engraving of the Hermitage, copied from the first and second editions, and forms the title-page of the "Natural History," as distinguished from the second half of the book, the "Antiquities." The Hermitage bears the words beneath it, "Where the Hermit hangs his straw-clad cell." Passages from Homer and Cicero then follow.

The next page is blank, and is followed on the next by:

Letter I of "The Natural History of Selborne."

Between pages 6 and 7 is inserted the plate of the fossil "Mytilus, Crista Galli," and by the note in the right-hand bottom corner ("volume I, page 12"), it is evident that the same plate has been used as in the 1802 edition. (See page 100).

Between pages 258 and 259 is a full-page engraving of the Stilt-Plover (*Charadrius himantopus*).

Title-page of the second part of the book, is found on page 303, as follows:—

"The Antiquities of Selborne, in the County of "Southampton," followed by the Seal of the Priory, and Virgil's lines "Juvat ire desertosque videre locos."

Letter I of the Antiquities commences on page 305.

A full-page engraving, "South View of Selborne Church," is inserted between pages 310 and 311; "P. Mazell, sculpt.; S. H. Grimm, del."

Another full-page engraving is inserted between pages 314 and 315, being "Copy of a Picture presented to the Church of Selborne, by Benjamin White, Esq."

Between pages 338 and 339 is an engraving, full page, of "Temple, in the parish of Selborne," and between pages 340 and 341, another of "The Pleystow, vulg. the Plestor."

On pages 418 and 419 are "More particulars respecting the Old Family Tortoise," etc.

"Appendix" commences on page 421. "A comparative view of the Naturalist's Calendar," etc.

"Observations on Various Parts of Nature. From Mr White's MSS. With remarks by Mr Markwick." A coloured illustration of "A Hybrid Bird," appears between pages 484 and 485.

"Meteorological Observations" commence on page 538.

"Observations" by Mr. Mitford appear on pages 543 to 548, and are signed "J. M., Benhall, Suffolk, 20th January, 1812."

"Summary of the Weather" commences on page 549.

"Poems" commence on page 561, and are—"The Invitation to Selborne;" "Selborne Hanger. A Winter Piece. To the Miss Batties;" "On the Rainbow;" "A Harvest Scene;" "On the dark, still, dry, warm weather, occasionally happening in the Winter Months."

Index, pages 573 to 587. List of plates, page 588.

No less than eight firms of publishers had a venture in this edition. The name of White still appears, but such familiar names as Longman, Bagster, and Hatchard, also have a place. It is the most complete of the older editions, and there are twelve high-class engravings, not the least interesting of which is a full-page copy of the beautiful triptych, which is now in the

Church, having been presented by Benjamin White in 1793.

This edition is sometimes know as Mitford's edition, in consequence of the notes on the Natural History on pages 543—548, by the Rev. J. Mitford, of Benhall, Suffolk. The short poem relating to the blossoming of the crocus is printed twice, viz. on page 236, at the end of Letter XLI to Barrington, and on page 570, with a number of other poems.

Some copies of this edition have been cut up into two volumes. A British Museum copy is in one volume, in a special binding, and bears the royal monogram "G.R." on the cover.

Second-hand catalogues: 12*s.* 6*d.*, 21*s.*, 18*s.*: 1888, 35*s.*, 15*s.*: 1890, £2 10*s.*

Auction Prices: 1893, 35*s.*: 1894, £2 6*s.* (with author's autograph letter to Sir J. Banks): 1909, 28*s*: 1908, 33*s.*: 1911, 26*s.*, 21*s.*: 1913, 21*s.*, 35*s.*: 1901, £1 10*s.*: 1902, £2 2*s.*: 1914, £1 13*s.*: 1915, 11*s.*, £5 5*s.* (with autographed bill), £1 13*s.*: 1917, 21*s.*: 1920, £2: 1922, £2 15*s.*: 1924, 26*s.*: 1926, 22*s.*: 1928, £1 4*s.*: 1929, £1 14*s.*: 1932, £2: 1933, £1 2*s.*

Mitford's Large-Paper Edition. There was also a Large-Paper edition published in two volumes price £5 5*s.* Fifty copies were printed. In these the picture of the Hybrid Bird is coloured, and the outline of the painting presented to the Church is coloured in imitation of the original picture.

Second-hand: 1889, £1 7*s.* 6*d.*: 1890, £2 10*s.*: 1896, £1 14*s.*: 1903, £9.

MITFORD'S ROYAL OCTAVO EDITION OF 1813. In two Volumes. Price, 16*s*. In pale green boards, with paper labels attached to the back of the volumes. On the label appears " 'White's Natural History.' New Edition, with four Plates. 2 Vols. 16*s*. boards. Vol. I (II)."

*Volume I.* Page i. "The Natural History of Selborne, &c., &c."

Page iii. Title-page. "The Natural History of Sel- "borne, by the late Rev. Gilbert White, A.M., Fellow "of Oriel College, Oxford. To which are added, The "Naturalist's Calendar, Miscellaneous Observations, "and Poems. A New Edition, with Engravings. In two "Volumes." Facing the title-page is an engraving of The Hermit's Cell. "Vol.I. London: Printed for White, "Cochrane, and Co.; Longman, Hurst, Rees, Orme, "and Brown; J. Mawman; S. Bagster, J. and A. Arch; "J. Hatchard; R. Baldwin; and T. Hamilton, 1813."

pp. v, vi, vii, viii, are devoted to John White's "Advertisement," dated "Fleet Street, 1802."

pp. 1 to 195, contain the Letters to Thomas Pennant. Opposite page 13 is an engraving of the fossil "*Mytilus, Crista Galli.*"

pp.196 to 351 (the last in the volume), contain the first twenty-nine Letters to Daines Barrington.

Page 351 ends as follows "End of Vol. I. T. Bensley, Printer, Bolt Court, Fleet Street, London."

*Volume II.* Title-page as in first volume. Facing it is an engraving of *Charadrius himantopus*, the stilt-plover.

pp. 1 to 160 contain Letters XXX to LXVI to Barrington.

pp. 161 to 195: The Naturalist's Calendar.

pp. 197 to 333: Observations on Various Parts of

Nature. Opposite page 213 is a picture of "A Hybrid Bird" in colours.

pp. 335 to 346: Poems. The poem, "Say, what impels," occurs here as well as on page 48.

pp. 347 to 364: Index.

Page 364: Part of Index, and "The End. T. Bensley, Printer, Bolt Court, Fleet Street, London."

Second-hand: 1910, 26*s*.: 1926, 22*s*.

EDITION OF 1822. Two volumes 8vo. Mitford's Notes. Includes "The Naturalist's Calendar," and "Observations," but not the "Antiquities."

*Volume I.* Title-page faced by coloured picture of the Hybrid Bird.

Title-page. "The Natural History of Selborne, by "the late Rev. Gilbert White, A.M., Fellow of Oriel "College, Oxford. To which are added, The Natural-"ist's Calendar, Miscellaneous Observations, and "Poems—A New Edition, with Engravings—In Two "Volumes—Volume I—London: Printed for J. & A. "Arch, Longman, Hurst, Rees, Orme, and Brown; "Lackington & Co.; J. Mawman, Baldwin, Cradock, "and Joy; J. Hatchard & Son; S. Bagster; Ogle, Dun-"can & Co.; W. Mason; J. Sheldon; R. Saunders; and "Hurst & Robinson. 1822."

On the back of title-page:—"Printed by T. C. Hansard, Peterboro' Court, Fleet Street, London."

On pages v—viii is the advertisement by "J. W." The whole reprinted from the 1802 edition and dated "J. W., Fleet Street, 1802."

Letters to Thomas Pennant comprise pages 1—195.

Letters I to XXIX of those to Daines Barrington, pages 196—350.

Opposite page 13 are three aspects of "Mytilus, Crista Galli."

The name of the printer again appears at the foot of the last printed page.

*Volume II.* The title-page is faced by an engraving of the Hermitage, by S. E. Grimm & W. Angus, entitled "Where the Hermit hangs his straw-clad cell."

Title-page, same as in volume I. Name of printer again appears on back of same.

The volume opens with Letter XXX to Daines Barrington, the series of letters being completed on page 160.

"More Particulars" of the Tortoise appears at the end of Letter L to Barrington. This is indexed as "More Remarks," etc.

"A Comparative View of the Naturalist's Calendar," etc., then follows on page 161, and is completed on page 195.

The "Observations on Various Parts of Nature. From Mr White's MSS., with remarks by Mr Markwick," are found on pages 197—306.

They are followed on pages 307—216, by the Rev. J. Mitford's notes, which appeared first in the 4to edition of 1813.

"Summary of the Weather" is found on pages 317—333.

From page 337 to page 346 are the usual Poems—"The Invitation to Selborne;" "Selborne Hanger. A Winter Piece. To the Miss Bxxxxxs;" "On the Rainbow;" "A Harvest Scene;" "On the early and late blowing of the vernal and autumnal crocus" (this also appears in the same volume at the end of Letter XLI to Daines Barrington, but with differences as to the words placed in italics); "On the dark, still, dry, warm weather," etc.

Index on pages 347—364, at the foot of the last page of which the printer's name appears.

In addition to the engraving facing the title-page in Volume II., there is one opposite page 43, of the Stilt-plover (*Charadrius himantopus*).

Second-hand: 9*s*.

EDITION OF 1825. In two volumes. The arrangement is again Dr. Aikin's, as in the edition of 1802, and contains "J. W's" preface, but the publishers are numerous and do not include J. White's name.

*Volume I.* First page is blank. Second page, "The Natural History of Selborne, &c., &c."

Opposite, and in front of title-page, is an engraving of the Hermitage, entitled "Where the Hermit hangs his straw-clad cell," reprinted from previous editions.

Title-page. "The Natural History of Selborne, by "the late Rev. Gilbert White, A.M., Fellow of Oriel "College, Oxford. To which are added, The Natural-"ist's Calendar, Miscellaneous Observations, and "Poems. A New Edition, with engravings. In two "volumes. Volume I. London: Printed for C. & J. "Rivington; J. & A. Arch; Longman, Hurst, Rees, "Orme, Brown, and Green; Harding, Triphook and "Lepard; Baldwin, Cradock and Joy; J. Hatchard & "Son; S. Bagster; G. B. Whittaker; James Duncan; W. "Mason; Saunders & Hodgson; and Hurst, Robinson "& Co. 1825."

The printer's name appears on the back of the title-page, as "T. C. Hansard, Pater-noster-row Press."

Pages vi, vii, viii. "Advertisement," the same as in the 1802 edition, dated "J.W., Fleet Street, 1802."

The Letters "To Thomas Pennant, Esq." extend from page 1 to page 195, inclusive. Those "To the Honourable Daines Barrington" extend from page 196 to page 351, the last page in volume I. At the foot of page 351, the printer's name is repeated.

*Volume II.* The title-page is faced by a coloured representation of "A Hybrid Bird."

"More particulars respecting the old family Tortoise" appear at the end of Letter L.

The Letters to Daines Barrington conclude on page 160.

"A comparative view of the Naturalist's Calendar," as kept at Selborne and Catsfield from 1768 to 1793, commences on page 161, and ends on page 195.

Page 197. "Observations on various parts of Nature. From Mr. White's MSS. With remarks by Mr. Markwick."

These are followed on pages 307 to 316 by the Rev. J. Mitford's notes in regard to various passages in the Letters, the only difference between this reprint and their original form in the 1813 edition, being that the reference numbers to the pages have been altered to suit the present edition. The notes are signed "J. M., Benhall, Suffolk, 20th January, 1812."

The "Summary of the Weather" follows, and concludes on page 333.

Page 335. "Poems." These include, "The Invitation to Selborne;" "Selborne Hanger. A Winter Piece. To the Miss Bxxxxxs;" "On the Rainbow;" "A Harvest

Scene;" "On the early and late blowing of the vernal and autumnal crocus;" "On the dark, still, dry, warm weather, occasionally happening in the winter months."

The index follows on pages 347 to 364. The name of the printer again appears at the foot of the last page, "T. C. Hansard, Pater-noster-row Press."

Second-hand: 8*s.* 6*d.*, 6*s.* 6*d.*, 7*s.* 6*d.*, 10*s.*, 13*s.* 6*d.*, 10*s.*: 1917, 21*s.*: 1925, £2 12*s.*: 1932, 12*s.* 6*d.*

Auction Price: 1928, £2 15*s.*

CONSTABLE'S EDITION OF 1829. Additions by Jardine. Vol. XLV of "Constable's Miscellany." 12mo.

A small engraving of the interior of a room, resembling one of the low-roofed rooms at "The Wakes," with a student seated in an armchair, faces the title-page.

Title-page. "The Natural History of Selborne. By "the late Rev. Gilbert White, A.M., Fellow of Oriel "College, Oxford. With additions by Sir William Jar-"dine, Bart. Edinburgh: Printed for Constable & Co., "and Hurst, Chance & Co., London, 1829."

On the back of title-page, "Edinburgh: Printed by Ballantyne and Company, Paul's Work, Canongate."

Pages v to x, "Introduction." Passage in Latin quoted from Scopoli. John White's biography of Gilbert is here quoted from the edition of 1802.

This Introduction is dated "Jardine Hall, August, 1829," and evidently emanates from Sir W. Jardine, although not signed. From observations he makes in his edition of 1851 he apparently considers the latter to be the first to which he gave his whole and undivided attention. (See page 139).

The next page is the title to "The Natural History of Selborne, Part I." "In a series of Letters addressed to Thomas Pennant, Esq." The next page is blank.

On page 3 commences "The Natural History of Selborne. Letter I."

On page 11, there being no picture of the "Mytilus," the passage announcing that the author had had the fossil engraved has been suppressed.

Page 133. Title to "The Natural History of Selborne Part II." "In a series of Letters addressed to the Honourable Daines Barrington."

Letter I commences on page 135.

The final Letter (LXVI) concludes on page 330.

The "Antiquities" do not appear in this edition, nor do the "Poems," "Observations on various parts of Nature," and the "Naturalist's Calendar."

There was another edition of Jardine in 1829, not connected with "Constable's Miscellany," but published by the same firm of publishers and printed by Andrew Shortrede, Thistle Lane, Edinburgh. Two other issues in 1829, with slightly different title-pages, are in the Selborne Society's library.

It is worthy of note that, in the edition of 1853 (National Illustrated Library), Jardine mentions that on the disarrangement of Mr. Constable's affairs, the Miscellany passed into other hands, and "this edition assumed several forms." I have seen editions of 1831, 1832, 1833 and 1836, but it would seem probable that there are other" forms" of this edition still extant.

Jardine's own copy (1829) with autograph signature and interleaved and annotated, was exhibited by the Selborne Society in 1913; also his copies of the 1830,

("Ewbank" misprinted "Ewebank") 1832, and 1836 (annotated and printed by Manning & Smithson, Ivy Lane, Paternoster Row, London) editions.
Second-hand: 3*s*. 6*d*. (1833 ed.).

JARDINE'S EDITION OF 1832.

Title-page. "The Natural History of Selborne, "By the late Rev. Gilbert White, A.M., Fellow of "Oriel College, Oxford. With additions by Sir Wil-"liam Jardine, Bart., F.R.S.E., F.L.S., M.W.S., Author "of 'Illustrations of Ornithology.' New Edition. "Printed for Whittaker, Treacher, & Co., London; "and Waugh & Innes, Edinburgh,—1832."

The Preface is dated "Jardine Hall, August, 1829."

The Index follows, to page xv.

*Part I;* the letters to Thomas Pennant extend to page 130. *Part II;* those addressed to Daines Barrington extend from pages 131 to 325.

At the foot of the last page, the printer's name appears, viz.: "London: Gilbert and Rivington, Printers, St. John's Square." Two pages of advertisements of books follow.

There are a few illustrations.

A few of White's "Observations on various parts of Nature," appear as footnotes.

JARDINE'S EDITION OF 1833.

The title-page is faced by an engraving of "The Village and Church of Selborne. J. Ewbank, Delt, W. Archibald, Sculpt." This is a reduced copy of the large folding view which appeared in the original edition of 1789.

Title-page. "The Natural History of Selborne;

"Observations on Various Parts of Nature; and the "Naturalist's Calendar. By the late Rev. Gilbert "White, M.A., Fellow of Oriel College, Oxford. "With additions by Sir William Jardine, Bart., "F.R.S.E., F.L.S., M.W.S., Author of 'Illustrations "of Ornithology.' New edition, with eighteen engrav- "ings, by Branston. Printed for Whittaker, Treacher, "& Co., London; and Waugh & Innes, Edinburgh. "1833."

The name of the printer appears on the back of the title-page; London: Gilbert & Rivington, Printers, St. John's Square."

"Introduction." Pages ii to vii, dated "Jardine Hall, August, 1829."

On page viii is a note dated at the head "Edinburgh, August, 1830," referring to the contents of this edition.

The following "Poems" then appear. "The invitation to Selborne;" "Selborne Hanger. A Winter Piece. To the Miss Bxxxxxs;" "On the Rainbow;" "A Harvest Scene;" "On the dark, still, dry, warm weather," etc.

Letter I appears on page 11. The Letters in this edition are arranged according to date, and not, as in the original edition, in two series. Letter CX, the last, thus closes on page 332.

Then follow "Observations on Various Parts of Nature, from Mr. White's MSS. With remarks by Mr. Markwick."

Mitford's notes, which appeared first in the edition of 1813, are reprinted here as footnotes to the "Observations."

"A Comparative View of the Naturalist's Calendar,"

etc., commences on page 407, and concludes on page 432.

Index, pages 433 to 440.

There are a few illustrations of birds in this edition.

Second-hand: 3*s*. 6*d*.

PROFESSOR RENNIE'S EDITION, 1833. This edition is bound in dark green cloth, and has a paper label affixed to the back with the following designation. "White's Natural History and Antiquities of Selborne, new edition, with Notes and Fine Cuts; price, 16s."

First page bears, besides "The Natural History," etc. the words "Embellished with Engravings—1833."

On second page is "Chiswick Press: C. Whittingham, College House."

Third page, "Natural History," etc. "A New Edi-"tion with Notes, by several eminent Naturalists. And "an enlargement of The Naturalist's Calendar." Then follows an engraving of "East Woodhay Warbler, W. "Herbert, del., London. Printed for J. and A. Arch: "Longman, Rees, Orme, Brown, Green, and Long-"man," etc., etc.

Amongst the naturalists who assisted Rennie were William Herbert (afterwards Dean of Manchester), and Robert Sweet.

Fourth page, blank. Fifth page, "Advertisement," being Gilbert White's Preface, continued on the sixth, seventh, and eight pages, dated Selborne, January 1st, 1788.

Ninth and tenth pages, "Biographical Records of the Author," being reproduced from the edition of 1802. On page 9 is "View of the residence, at Selborne, of the late Rev. Gilbert White."

Eleventh page, "Preface to this Edition;" continued on twelfth page, and signed "J. R., Lee, Kent, 16th June, 1832."

Thirteenth page, "The Natural History," etc., followed by a view of "The Village." Then six lines by White, commencing "See, Selborne spreads her boldest beauties round." Letter I, "To Thomas Pennant, Esquire," then commences at the bottom of this page.

Page 3 of letters has an engraving of "The Hanger," by Nesbit; page 6, one of "The Plestor;" page 9, The Fossil Oysters; page 13, at the head of Letter V, "The Little Bridge;" and as a tail-piece, "Approach to the Village of Alton." Other illustrations are:—

The Harvest Mouse, page 38. Jackson sc.

Wood-Wren, page 53. W. Herbert delt.

Wood-Wren, page 56. W. Herbert delt.

An unnamed Nest, etc., page 58. H. W. Herbert delt.

Nest and Eggs of *Silvia Silviella*, page 126. H. W. Herbert delt.

On page 141 commences Letter I to the Honourable Daines Barrington.

"More particulars respecting the Old Family Tortoise," follow Letter L on page 299.

"Observations on Vegetables," commence on page 383. "Poems" on page 401.

"Summary of The Weather" on page 411.

"A Comparative View of the Naturalist's Calendar as kept at Selborne, in Hampshire," etc., appears on page 419.

"Calendar of British Zoology," by James Rennie, M.A., commences on page 435.

"The Antiquities of Selborne, in the county of Southampton," with the Seal of the Priory, and ". . . . Juvat ire . . . ., Desertosque videre locos . . . . Virgil," appear on page 449.

An engraving by Jackson, "South View of Selborne Church," appears on page 456.

On page 464 is "The Vicarage House." Jackson.

On page 482 is "Temple in the Parish of Selborne." Jackson.

On page 555 is "Index to the Natural History," and on page 561, "Index to The Antiquities."

Total, 562 pages, followed by two pages of "Works recently published."

This is one of the few editions which contain the whole of the "Antiquities," as well as the "Naturalist's Calendar," the "Observations," and the Poems. As a head-piece to the short biography of White, which first appeared from J. White's pen in the edition of 1802, and which is reprinted in this edition, there is an authentic view, from the garden, of "The Wakes," as the residence of the author at Selborne was, and is still known. There is a similar illustration in the 1813 edition, and these both represent the condition of the building as it actually appeared in White's time. Much as it has already been altered since it passed out of the White family a hundred years ago, there is yet enough in the old parts of the house to recognise the home of the naturalist. In the illustration referred to, the additional room which White caused to be built, and to which he constantly refers in his correspondence, is seen on the left hand of the picture.

Second-hand: 3*s.* 6*d.*, 5*s.*, 7*s.* 6*d.*, 4*s.* 6*d.*, 10*s.*: 1887, 45*s.*

# CHAPTER VIII

## EDITIONS OF LADY DOVER, S.P.C.K., AND THOMAS BROWN

LADY DOVER'S DUODECIMO EDITION OF 1833. Published by N. Hailes. 316 pages. First issue of Lady Dover's editions. See S.P.C.K. edition of 1842.

LADY DOVER'S EDITION OF 1840. Published by Thomas Tegg, No. 73 Cheapside. With 53 engravings by J. Thompson. Printed by Balne Brothers, 38 Gracechurch Street, London. Vignette of "The Wakes" on title-page.

S.P.C.K. EDITION OF 1842. "Arranged for Young Persons." 6*s*. 8vo.

This edition is a subsequent issue of Lady Dover's anonymous and bowdlerised edition of 1833 (Newton).

The printer's name appears opposite the title-page as "London: R. Clay, Printer, Bread Street Hill."

The "Dedication" is found on pages v and vi and is "To H.A.E.," (i.e. H. Agar-Ellis) and is signed "Your affectionate Mother. May 25, 1833."

The edition was dedicated to Lady Dover's son, afterwards Lord Clifden.

On page vii is an illustration of Gilbert White's house, followed by an "Account of the Rev. Gilbert White." This is his nephew's biography from the edition of 1802.

An illustration, the "Village of Selborne," appears on page 2 of the Letters. There are numerous illustrations of birds, etc. "Hawkley Slip" appears on page 265.

The last page of the Letters is page 325, this edition including the natural history letters only, with, in suitable places, a few extracts from the Antiquities, subjoined as notes.

Index appears on pages 327–8.

This edition is stated by Newton to have been the foundation of a large number of American reprints from 1841 onward. The S.P.C.K. informed me that editions of 1,000 each were issued in 1860, 1862, 1871, 1875, 1877, 1880, 1883, 1888 and 1893. Also in 1908.

Second-hand: (no date) 2*s*.

American Edition of 1843. Antiquities, Observations, and Naturalist's Calendar, omitted. First issued in 1841, and founded on Lady Dover's incomplete edition (see S.P.C.K. edition of 1842).

The title-page runs thus:—"The Natural History of "Selborne. By the Rev. Gilbert White, A.M., Fellow "of Oriel College, Oxford. New York: Harper & "Brothers, Cliff Street—1843;" and bears an illustration, reduced from Rennie's edition of 1833, of the back view of "The Wakes," Gilbert White's residence at Selborne.

On the back of the title-page is the legend; "Entered, according to Act of Congress, in the year 1841, by Harper & Brothers, In the Clerk's Office of the Southern District of New York."

"Account of the Rev. Gilbert White" follows, being reprinted from the edition of 1802, but with no acknowledgment of its source, nor with the initials

"J.W." which appeared beneath it in that edition.

The Index then appears, at the commencement of the book, on pages ix and x.

On page xi is a "List of engravings," fifty-two in number. The engravings themselves, however, bear no inscriptions to show what they represent, and that of *Himantopus* on page 286, would scarcely be recognised as such, being in a sitting posture.

From the heading above Letter I to Pennant, it appers that the edition has been "arranged for young persons." The first series ends on page 143.

Letters to Barrington occupy pages 145-335, on the last page of which "Aikin" is misspelt "Aiken."

"Catalogue of books" appears on the back of the last page.

The numbers of the letters do not in some cases correspond with the numbers prefixed to them in the original edition of 1789.

"More particulars respecting the old Family Tortoise," is printed on pages 290-291, Letter XLVI.

There were other American reprints of this edition.

Captain Thomas Brown's Edition of 1833.

This is the first of several editions in 12mo. with notes by Capt. Thomas Brown, F.L.S. The notes are upon a somewhat extensive scale, almost rivalling in extent the original matter. There are a few mediocre illustrations. The two series of letters have been arranged in strict chronological order, instead of in the manner of the original edition, which kept the letters to Pennant distinct from those to Barrington. It is noticeable that the successive editions are by various publishers from stereotyped plates.

Volume I of "The British Library" (Newton).

Facing the title-page is a small illustration, on a full page, of "Selborne Church, from the north."

Title-page. "The Natural History of Selborne, "Observations on various parts of Nature; and the "Naturalist's Calendar. By the late Rev. Gilbert "White, A.M., Fellow of Oriel College, Oxford. "With notes by Captain Thomas Brown, F.L.S., "M.K.S., &c., President of the Royal Physical Society, "Edinburgh: Published, for the proprietors, By James "Chambers, Edinburgh; W. Orr, London: and W. "Curry, Jun. & Co., Dublin, MDCCCXXXIII."

On the back of title-page is "Edinburgh: Printed by Andrew Shortrede, Thistle Lane."

The Preface commences on page iii, and opens with a reprint of John White's biographical sketch of the author. The preface is dated Edinburgh, January 25th, 1833, and was written sufficiently near to the time of White's life at Selborne, to cause the writer to distinguish the naturalist by the title of "Mr." Pages iv, v and vi are also devoted to the preface.

On pages vii to xii are the following poems:—"The Invitation to Selborne;" "Selborne Hanger; A Winter Piece.—To the Miss Bxxxxxs;" "On the Rainbow;" "A Harvest Scene;" "On the dark, still, dry, warm weather, occasionally happening in the Winter Months."

Between the poems and the letters are placed seven full-page illustrations, viz., Cowthorpe Oak, at Wetherby; A Hybrid Pheasant; the Parrot Crossbill; Domestic Hen with Male Plumage; the Moose-deer, or elk; the

Bohemian Wax-wing; and the Hoopoe. These are differently inserted in different editions of Captain Brown.

Letter I to Thomas Pennant follows. As the letters in this edition are arranged in the order of their dates, the last letter is numbered CX, and concludes on page 282.

In a footnote to page 4, Sir Adam Gurdon's name is spelt "Gordon," and this is repeated in all of Brown's editions.

The letters are followed by "Observations on various parts of Nature, from Mr. White's MSS., with remarks by Mr. Markwick and the Editor." "Observations on Quadrupeds," commences on page 283.

Pages 331 to 336 contain the "Summary of the Weather."

On page 337 commences "A Comparative View of the Naturalist's Calendar," as kept at Selborne and Catsfield from 1768 to 1793. This continues to and includes page 348.

On pages 349 to 356 there is a full Index. The name of the printer appears again at the foot of the last page, "Edinburgh: Printed by Andrew Shortrede, Thistle Lane."

The "Antiquities" do not appear in this edition.

This volume is bound in boards with figured cloth, the title and name of editor appearing at the back of the book on an adhesive label.

There were two other printings in this year (1833) with slightly differing title-pages.

CAPTAIN THOMAS BROWN'S SECOND EDITION OF 1834.

The second edition varies in some respect from the first edition. The publishers are given on the title-page as, "Edinburgh, Frazer & Co., North Bridge: Orr & "Smith, Paternoster Row, London: W. Curry, Jun. "& Co., Dublin, MDCCCXXXIV."

Inserted before p. 1 is a slip entitled "Directions for placing the Plates." This seems to show that their manner of insertion in the first edition was not the desired one.

The name of the book is stamped in gold on the back of the green cover, "White's Natural History of Selborne."

Second-hand: *2s.*, *3s.*

A Third Edition was published in 1834 by Fraser, Smith Elder and Curry.

CAPTAIN THOMAS BROWN'S FOURTH EDITION OF 1835.

Published by Fraser, Orr & Smith, and Curry.

Title-page. "The Natural History of Selborne; with "Observations on Various Parts of Nature, and the "Naturalist's Calendar. By the Late Rev. Gilbert White, "A.M., Fellow of Oriel College, Oxford. Edited by "Captain Thomas Brown, F.L.S., &c. Fourth Edition. "Edinburgh: Published by Fraser & Co.; Orr & Smith, "Pater-noster Row, London; W. Curry, Jun. & Co., "Dublin; MDCCCXXXV."

For notes on remainder of this edition, see the edition of 1833.

A further issue was made in 1835 by F. J. Mason, 444 West Strand.

CAPTAIN THOMAS BROWN'S FIFTH EDITION OF 1835. Published by Tegg & Son, R. Griffin & Co., and Tegg, Wise & Co. Paper label on back of brown cardboard cover.

Title-page. "The Natural History of Selborne; with "Observations on Various Parts of Nature, and the "Naturalist's Calendar. By the late Rev. Gilbert White, "A.M., Fellow of Oriel College, Oxford. Edited by "Captain Thomas Brown, F.L.S., etc. Fifth Edition. "London: Thomas Tegg & Son, Cheapside; R. Griffin "& Co., Glasgow; Tegg, Wise & Co., Dublin, 1835."

In this edition the eight full-page illustrations are all together, before Letter I to Pennant.

For notes on the remainder of this edition, see the edition of 1833.

CAPTAIN THOMAS BROWN'S SIXTH AND SEVENTH EDITIONS OF 1836. Published by Fraser & Washbourne.

Title-page. "The Natural History of Selborne; with "Observations on Various Parts of Nature, and the "Naturalist's Calendar. By the late Rev. Gilbert White, "A.M., Fellow of Oriel College, Oxford. With exten-"sive additions by Captain Thomas Brown, F.L.S. etc. "Sixth (or Seventh) Edition. Edinburgh: Fraser & Co., "54 North Bridge; Henry Washbourne, London, "MDCCCXXXVI."

For notes of remainder of this edition, see the edition of 1833.

A further issue was made in 1836 by Allan Bell & Co. and Washbourne.

In 1838 a similar edition was issued, but strangely enough, this is also called the "Seventh Edition," and the date on the title-page is in arabic numerals.

Captain Thomas Brown's Eighth Editions of 1840 and 1845. Published by John Chidley.

Facing the title-page is an illustration of Selborne Church, from the North.

Title-page. "The Natural History of Selborne; with "Observations on Various Parts of Nature, and the "Naturalist's Calendar. By the late Rev. Gilbert White, "A.M., Fellow of Oriel College, Oxford. With exten-"sive additions by Captain Thomas Brown, F.L.S., etc. "Illustrated with eight engravings. Eighth Edition. "London: John Chidley, 123 Aldersgate Street. "MDCCCXLV."

For notes on the remaining portion of this edition see edition of 1833.

This edition (eighth) has no Index. It is not a little remarkable that after searching for a copy of this edition for years without success, I found no less than five new copies of it on a coster's barrow in Farringdon Street (London) in 1905.

Another so-called eighth edition was issued in 1843 by I. J. Chidley.

Captain Thomas Brown's Tenth Edition of 1853. Publisher, R. Griffin & Co.

Captain Thomas Brown's Editions of 1856 and 1857. These further editions in 12mo. were published by Blackwood, price 5*s*.

THE GOLDEN LIBRARY, 1875. WITH THOMAS BROWN'S NOTES. Chatto & Windus.

Title-page. "The Natural History of Selborne. With "Observations on various parts of Nature, and The "Naturalist's Calendar. By the Rev. Gilbert White, "M.A., edited, with additions, by Thomas Brown, "F.L.S. Chatto & Windus, Piccadilly, 1875."

Preface on pages iii and iv. This is John White's advertisement of 1802, with three paragraphs from Brown's first edition of 1833.

The Letters are numbered chronologically, and not as in the original edition.

There are a few illustrations to the "Observations on Insects."

AN EDITION WITH NOTES BY THOMAS BROWN, 1880. Published by Chatto & Windus, 8vo. Title in gilt lettering on back.

Title-page. "The Natural History of Selborne with "Observations on various parts of Nature and the "Naturalist's Calendar, By the Rev. Gilbert White, "M.A. Edited by Thomas Brown, F.L.S." Publisher's Coat of Arms. "A New Edition, London, Chatto and "Windus, Piccadilly, 1890."

Page ii. "Ballantyne Press, Ballantyne, Hanson, & Co., Edinburgh and London."

Pages iii and iv. Preface. Same as given by John White in Edition of 1802, followed by three original paragraphs, commencing "Gilbert White is principally known to the world," etc. Mention is made that a translation was made in Berlin in 1792.

On the fifth page from the title-page commences the first letter to Thomas Pennant, Esq.

Some lines are omitted at the end of Letter III relative to an engraving of the oyster referred to, since in this edition there are no illustrations.

The letters are inserted in this Edition in the order of their dates, and numbered accordingly, whether they be to Thomas Pennant or Daines Barrington.

Page 282, the last page of the Letters.

Page 283. "Observations on various parts of Nature, from Mr. White's MSS., with remarks by Mr. Markwick and the Editor." "Observations on Quadrupeds."

Page 285. "Observations on Birds."

Page 306. "Observations on Insects and Vermes."

Small illustrations on pages 310, 311, 312, 314, and 315.

Page 320. "Observations on Vegetables."

Page 327. "Meteorological Observations."

Page 331. "Summary of the Weather."

Page 336. "A comparative View of the Naturalist's Calendar..... From the year 1768 to the year 1793."

Page 348, "In sese vertitur annus."

Note.—The "Antiquities of Selborne" are omitted in this edition, as they are said in the preface to offer "few attractions to the general reader." A similar note appears in other editions of Captain Brown.

Edition of 1890 with Thomas Brown's Notes. Chatto and Windus. 348 pages. Printer, Ballantyne, Hanson and Co. With the Observations and the Naturalist's Calendar.

## CHAPTER IX

## EDITIONS OF BLYTH, BENNETT, AND JENYNS

Blyth's Edition of 1836. 8vo. The details of the 1850 edition (see following) refer also to this one, with the following exceptions. On the title-page is a figure of three miniature waterfalls. The publishers are "Orr and Smith, Paternoster Row," and the date is MDCCCXXXVI.

Pages i to iii give the "Advertisement" by Edward Blyth and pages (repeated) i to xix give Mudie's description of Selborne. Opposite the title is a map of the "Environs of Selborne, drawn and engraved by I. Dower, Pentonville, London."

The printer's name appears after the Index at the end of the book, as "Thoms, Printer, Warwick Square."

Each page is not framed in black lines as in the editions of 1850 and 1858.

Second-hand: *2s.*, *4s.*

Blyth's Edition of 1850. One volume, 8vo., gilt-edged. Map of the "Environs of Selborne," opposite title-page in some copies. Each page is framed in double black lines.

Title-page. "The Natural History of Selborne, "with its Antiquities; Naturalist's Calendar, &c., by "the Rev. Gilbert White, A.M. A New Edition, with "notes by Edward Blyth. To which is added a description of the village and neighbourhood, written on the "spot for this edition by the late Robert Mudie. "London: Published by Wm. S. Orr & Co., Amen "Corner, Paternoster Row, MDCCCL."

Back of title-page, "London: Printed by Stewart & Murray, Old Bailey."

Page ii. "Advertisement." The first four paragraphs of this preface consist of the author's original advertisement or preface, the remainder being by Mr. Blyth. Reference is made at its conclusion to the "interesting account of Selborne" written for this edition by Mr. Mudie, who visited Selborne in 1836. The preface is dated "Lower Tooting, Nov., 1836."

Page iv. A quotation from Cowper.

(Page v.) "The Natural History," Seal of the Priory, and "Antiquities of Selborne." This page is differently placed in some copies.

Page vii is page i of Mudie's description. On the same page is a woodcut of the gravestone, bearing the inscription "G.W., 26th June, 1793."

On page v is an illustration of a country lane. On page vi is a woodcut evidently of Selborne, showing the Hanger and Nore Hill, but the proportions are incorrect. On page xi is a woodcut of a village street.

John White's short biography appears on page xvii (misprinted xvi) in Mudie's description, which is brought to a close on page xix of the same. The first letter of the Natural History then follows.

There are numerous small woodcuts scattered through the volume, mostly of birds.

Three views of the fossil oyster appear on page 6; one of the Red Deer on page 15. A cut of Stonehenge occurs on page 69.

The first Letter to the Hon. Daines Barrington commences on page 117, the letters to this gentleman being rightly kept distinct from those to Mr. Thos. Pennant.

On pages 135 and 136 the letter supposed to have been written by Timothy, the tortoise, to Miss Mulso, is printed (see page 78).

A woodcut of the Peregrine Falcon occurs on page 241.

The last letter (No. LXVI) is dated at the foot "Selborne, June 25, 1787," and is followed on page 260 by "More particulars respecting the Old Family Tortoise, omitted in the Natural History."

On page 261 commences a series of "Observations on Various Parts of Nature." These are followed on page 303 by "Meteorological Observations," on page 306 by "A summary of the Weather, from the year 1768 to 1793."

On page 313 commences "A comparative view of the Calendar of Nature," as kept by White and Markwick from 1768 to 1793.

"The Antiquities of Selborne" begin on page 323.

There is a woodcut of "Well-head" on page 325, although it is impossible to judge of the position of the spring in the picture.

On page 327 is a good representation of "Selborne Church," from the east. On page 349 an illustration of "The Plestor" appears.

The "Antiquities" end on page 405, being followed by poetry.

The poetical pieces which are printed are "The Invitation to Selborne;" "Selborne Hanger. A Winter Piece," dated November 1, 1763; "On the Rainbow," dated February 13, 1783; "A Harvest Scene;" "On the early and late Blowing of the Vernal and Autumnal Crocus;" "On the dark, still, dry, warm weather, occasionally happening in the winter months."

"Index to the Natural History," page 413.

"Index to the Antiquities," page 418.

"Stewart & Murray, Old Bailey."

A page of advertisements of "Works on Natural Science" then follows. A further edition appeared in 1851.

Second-hand: 3*s.* 6*d.*

BLYTH'S EDITION OF 1858.

In most respects this edition is identical with that of 1850. The differences are as follows:

Title-page. The publisher's name appears as "Thomas Nelson & Sons, London, Edinburgh and New York. MDCCCLVIII." On back of the title-page, the printer is given as "Thomas Harrild, Printer, Salisbury Square, Fleet Street."

The map of the environs of Selborne is omitted.

Each page is framed in a double black line, as in the editions of 1850.

BENNETT'S EDITION OF 1837. "Embellished with Engravings."

Title-page. "The Natural History and Antiquities of "Selborne. By the Rev. Gilbert White, M.A. With the "Naturalist's Calendar; and Miscellaneous Observa-"tions. Extracted from his papers." There is here a small engraving of the Hermitage, by Thompson, although not named. "A New Edition, with Notes, by "Edward Turner Bennett, Esq., F.L.S., &c., Secretary "of the Zoological Society; and others. London: "Printed for J. and A. Arch; Longman & Co.; Baldwin "& Cradock; Hatchard & Son; R. Scholey; J. G. and "F. Rivington; Whittaker & Co.; J. Duncan; J. Capes; "W. Mason; E. Hodgson; J. Bain; W. J. and J. May-"nard; J. Bohn; J. Vanvoorst; and Houlston & Son.'

Page v and vi are devoted to Gilbert White's "Advertisement."

Page vii. "Biographical Records of the Author." This is John White's short biography. It is signed in this edition, "I.W."

"Preface to the present Edition" extends from page ix to xxiii, and is written by a brother of the editor, the latter having predeceased publication of the edition. It is signed "I.J.B., Bulstrode Street, Sept. 15th, 1836."

The List of Illustrations, of which there appear to be forty-six, is printed on page xxiii.

Letter I to Thomas Pennant follows, and is headed by an illustration of "Selborne, from Dorton," reprinted from Rennie's edition of 1833, where it is called "The Village."

On page 16 is an illustration of the "Hollow Lane, and Bridge, near Norton." This is reprinted from page 13 of Rennie's edition, where it is called "The Little Bridge."

"Gilbert White's Tortoise" is shown on page 361.

Page 401 is a title-page to "A Naturalist's Calendar, "with Observations in various branches of Natural His-"tory. Extracted from the papers of the Rev. Gilbert "White; by John Aikin, M.D. With remarks by Mr. "Markwick and others." This is followed on page 403 by Dr. Aikin's "Advertisement," dated "J. Aikin, London, Jan. 1, 1795." An appropriate "Note" by "E.T.B." appears on this and the next page, concerning Dr. Aikin and his work.

"The Naturalists' Calendar," etc., commences on page 405. This is prefaced on page 407 by a quotation from Dr. Aikin's preface in the volume of 1795, and a paragraph in reference to the parallel calendar of Mr. Markwick, from the second edition of 1802; also by a note by "E.T.B." as to Mr. Markwick himself.

"A Comparative View," etc., commences on page 409.

"Observations of Various Parts of Nature, from Mr. "White's MSS., with remarks, by Mr. Markwick and others," commences on page 425. There are numerous illustrations.

"Summary of the Weather," commences on page 497.

"Poems, selected from the MSS. of the Rev. Gilbert White," commence on page 503. These include "The Invitation to Selborne;" "Selborne Hanger, A Winter Piece. To the Miss Batties;" "On the Rainbow;" "A Harvest Scene;" "On the dark, still, dry, warm Weather."

"The Antiquities of Selborne, in the County of Southampton," with the Seal of the Priory, and a quotation from Virgil, appear on page 513.

Illustrations appear of "South View of Selborne Church," on page 521; "The Vicarage House," on page 530; "Temple in the parish of Selborne," on page 550; "The Plestor," on page 552; these are all reprinted from Rennie's edition.

An illustration appears on page 627 of the "Way

leading to Gracious Street." This is entitled in Rennie's edition, "The Hanger."

Index appears on pages 631 to 640 inclusive, at the foot of the last page of which is the printer's name, "Chiswick: Printed by C. Whittingham."

The notes to this edition are very extensive. The edition is similar to Rennie's of 1833, and is published by the same publishers. Some of Rennie's illustrations are reprinted.

The preface to this edition, by I. J.B., contains a very good summary of the various editions previously published.

"More particulars respecting the old family tortoise" appears on page 359, at the end of Letter L to Daines Barrington.

Second-hand: 18*s.*, 9*s.* 6*d.*, 10*s.*, 15*s.*, 6*s.*, 16*s.*: 1894, 6*s.*: 1897, 36*s.*: 1932, 6*s.* 6*d.*

JENYNS' EDITION OF 1843. Van Voorst, publisher. This edition omits the Antiquities, the Poems, the Observations on Various Parts of Nature, and the Naturalist's Calendar. It is bound in boards with a figured cloth, somewhat in old-fashioned prayer-book style; 8vo.

First page, "The Natural History of Selborne." Second page, a crest and initials with the motto, "Labore et Constantia." Published at 7*s.* 6*d.*

Title-page, on third page. "The Natural History of Selborne. By the late Rev. Gilbert White, M.A." The words "Wisdom of God in Creation" then appear in a garter, surrounding a portrait bust of a divine. Then follows a Latin quotation from "Scopoli Ann. Hist. Nat." "A new edition, with notes by the Rev. Leonard "Jenyns, M.A., F.L.S., etc. London: John Van Voorst, "Paternoster Row, MDCCCXLIII." Back of title-page, blank.

Preface, pages v to xiii inclusive. This ends, as usual, with John White's biographical record of his uncle from the edition of 1802, and is dated "L.J., Swaffham Bulbeck, December, 1842."

Pages xv and xvi contain the List of Illustrations, twenty-six in number.

Letter I to Thomas Pennant, Esq., then follows, the series to this gentleman ending on page 159. Part II. "Letters addressed to the Hon. Daines Barrington" commences on page 161, this series ending on page 384.

The Index is found on pages 385 to 398, and at the foot of page 398 is the printer's name, etc., "London: Printed by S. and J. Bentley, Wilson, and Fley, Bangor House, Shoe Lane." Two pages of advertisements of books follow.

Some copies of this edition were re-issued in 1904, in a modern green cloth binding.

Second-hand: 4*s*., 1*s*. 6*d*.

# CHAPTER X

## EDITIONS OF JESSE AND JARDINE

JESSE AND JARDINE'S EDITION OF 1851. Bohn's Illustrated Library (First issue was in 1849). An engraving, "Selborne," appears opposite the title-page. The names "Davis" and "Hinchliff" appear at the bottom corners of the engraving. Red Cloth: 5*s*. With coloured illustrations 7*s*. 6*d*.

Title-page. "The Natural History of Selborne; "with Observations on various parts of Nature; and "The Naturalist's Calendar. By the late Rev. Gilbert "White, A.M., Fellow of Oriel College, Oxford. With "Addition and Supplementary Notes by Sir William "Jardine, Bart., F.R.S.E., F.L.S., M.W.S."

"Edited, with further illustrations, a biographical "sketch of the author, and a complete index by Edward "Jesse, Esq., Author of 'Gleanings in Natural History', "&c., &c. With forty engravings."

"London: Henry G. Bohn, York Street, Covent Garden, MDCCCLI."

Back of title-page, "London: Bradbury & Evans, Printers, Whitefriars."

Pages iii and iv. List of Illustrations. These are forty in number.

Page v. "A Short Biography of the Rev. Gilbert White." This contains a review of White's book from the pen of his brother Thomas, in the *Gentleman's Magazine* in 1789 (see page 31). Also the interesting letter by White, supposed to have been written by Timothy the tortoise to Miss Mulso.

This edition commences with "Poems, selected from the Manuscripts of the Rev. Gilbert White," namely, "Invitation to Selborne;" "Selborne Hanger. A Winter Piece. To the Miss Bxxxxxs;" "On the Rainbow," with a quotation from Eccles., xliii, 11, to which a note is appended stating that this, and "A Harvest Scene," which follows, "were published in the *Gentleman's* "*Magazine* for 1783, page 955, as imitations of an old "poet." Page 8 contains the poem "On the dark, still, dry, warm weather, Occasionally happening in the Winter Months."

Page 11. "The Natural History of Selborne. In a "Series of Letters addressed to Thomas Pennant, Esq,. "and the Hon. Daines Barrington."

Letter I commences at page 13, "To Thomas Pennant, Esq."

Principal Illustrations. The Raven, opposite 19; Black Grouse, opposite 27; The Fallow Deer, opposite 35; and numerous other engravings of birds and other animals.

The letters are arranged chronologically as to dates, those to Pennant not being kept separate from those to Barrington.

Page 303. "Observations on Various Parts of Nature from Mr. White's MSS. With remarks by Mr. Markwick."

Page 305. "Observations on Birds."

Page 356. "Meteorological Observations."

Page 359. "Summary of the Weather."

Page 367. "A Comparative View of the Naturalist's Calendar," by White, at Selborne, and Markwick, at Catsfield, from 1768 to 1793.

Page 393. "Supplementary Notes to White's Natural History of Selborne, by Sir W. Jardine, Bart, &c., &c., &c."

In the remarks on page 395, dated "Jardine Hall, 14th November, 1850," Sir William states with some warmth of feeling that although many editions had been published bearing his name, the only one with which he had any previous connection was that of 1829, published in "Constable's Miscellany."

Page 407. "Supplementary Notes by, or communicated to, the Editor." The majority of these notes appear to be by R. C. Norman.

Index, pages 412-416.

"Bradbury & Evans, Printers, Whitefriars."

A copy of this edition with coloured illustrations and in red cloth, in my possession, bears the book-plate of the late Professor Thomas Bell, who edited the work in 1877.

Second-hand: *2s.*, *4s. 6d.*, *2s. 6d.*: 1932, *4s. 6d.*

JARDINE'S EDITION OF 1853. National Illustrated Library. One volume. *2s. 6d.*, cloth, *7s. 6d.* morocco. Published by Nathaniel Cooke. There were other styles of binding at intermediate prices.

The frontispiece to this edition consists of a full-page illustration, entitled "General View of Selborne." This is faced by a map, "Environs of Selborne," not, however, reliable at the present day. This map is in some copies placed a few pages later.

On the next leaf is an illustration of an avenue of trees, the church tower in the distance, whilst in the foreground is a quaint old gentleman touching a small snake with a stick. Overhead among the branches is the inscription "The Natural History of Selborne by Gilbert White." The word "Selborne" is noticeable from the fact the letter N is reversed.

The next leaf contains the title-page: "The Natural "History and Antiquities of Selborne, with Observa- "tions on various parts of Nature, and The Naturalist's "Calendar. By the late Rev. Gilbert White, A.M. A "New Edition. Edited, with notes, by Sir William "Jardine, Bart., F.R.S.E., F.L.S., &c. Completely illus- "trated with about seventy engravings, comprising "subjects from Natural History, and Views of Selborne, "its Vicinity and Antiquities, sketched from Nature "expressly for this edition. London: Nathaniel Cooke, "Milford House, Strand. 1853."

Pages iii and iv contain the "Advertisement to original edition," signed "Gil: White, Selborne, January 1st, 1788.

Page v is headed by a half-page illustration of the "Village Street—White's house," followed by "Introductory Observations," and these are continued on pages vi to xiv. Jardine here states that, with the exception of some notes in the edition in "Constable's Miscellany" of 1829, and the volume in Bohn's Standard Library, he had no connection with any previous edition, and it is apparent that this edition (1853) is the first which is, in its entirety, Jardine's edition.

Dr. Aikin's name is misspelt "Aitkin." Gilbert White is called "Vicar of Selborne," which he never was. John White's sketch of Gilbert White's life in the 1802 edition, is quoted and reference is made to the lives of four of Gilbert's brothers, John, Thomas, Benjamin, and Harry.

On page xiii is a "Back view of White's House," a

good illustration, as indeed are all in this edition.

On page xiv is a cut of "White's Sun-dial," and also one of "White's Tombstone in Churchyard."

Page xv. "Contents." Pages xvii and xviii. "List of Illustrations," also containing list of plates. Here is inserted a full-page plate, "Old View of Selborne," and this is followed by Letter I of the Natural History, to Pennant. Letter I to Barrington commences on page 90.

The principal illustrations to the Natural History are:—Well-head, page 3; Ostrea carinata, p.7; Hollow Lane, p.10; Rocky Hollow Lane, p.13; Wolmer Forest, p.14; Rock Swallow, p.70; Great Bat and Honey Buzzard, between pp. 86-87; Rush-holder, p.142; Viper's Head and Tortoise, between pp.148-9; Rivulet in Short Lithe, p.172.

"The Antiquities of Selborne," page 201.

Illustrations to the "Antiquities." "Selborne Church and Vicarage," a good illustration, showing the yew-tree, p.204; Village Pleystow, p.221; Iron Key of ancient construction, and Steel Hinge, with griffin on it, p.240; Old Coins, p.241; Encaustic Tiles (8), p.245; Stone Coffin, p.247; Priory Seal, p.265.

"More particulars respecting the Old Family Tortoise" appear on pages 264 and 265.

"Observations on Various Parts of Nature," from White's MSS. with remarks by Mr. Markwick, page 267.

"Advertisement," page 268, quoted from edition of 1802. Illustrations of "Peregrine Falcon" and "Hybrid Pheasant," between pages 274-5.

Poem, "On the dark, still, dry, warm weather, occasionally happening in the Winter Months," p. 305.

Summary of the Weather, pages 307-312.

"A Comparative View of the Naturalist's Calendar," by White and Markwick, pages 313-330.

On pages 315-318 there is an "Introduction to Naturalist's Calendar," by Jardine, who reprints here some remarks from an ornithological calendar which he prepared in 1849.

The Poems then follow on pages 333-337. They are "The Invitation to Selborne;" "Selborne Hanger. A Winter Piece;" "On the Rainbow;" "A Harvest Scene."

Index, pages 339-342. "The End. Bradbury and Evans, Printers, Whitefriars."

This is one of the best of the older editions. Two errors appear in the note to page 163. Original edition was 1789, not 1780. Also the last part of Letter XLI to Barrington referred to in the same note was first put in its correct place in Aikin's edition of 1802, and followed as a matter of course in Mitford's 4to edition of 1813. But neither of these printed it for the first time. It is found, out of place, in the original edition, on pages 429 and 430. (See page 95.)

Another curious error crept into this edition. In Letter XLVII to Barrington, White speaks of the house cricket, as being of "a thirsty race," and later in the same Letter that they "are not only very thirsty, but very voracious." But Jardine prints it so as to say that they are "a thrifty race," and "not only thrifty, but very voracious." No one can imagine that the cricket is addicted to thrift. The mistake no doubt arose from the fact that in the original edition the long "s" is used, which is much like an "f"; then with the transposition of two other letters, we get the word "thrifty." Later editions were apparently founded on this edition of Jardine's, for the same error is found in the following: Prof. Morley's (Cassell), 1887; Christopher Davies' (Warne's Chandos Classics), 1879 and 1890: Blackie's, 1895; and in Library of Famous Books (no date); and Mosley's "Naturalist's Edition" (Elliot Stock), 1905.

Re-issues were made in 1879, 1882, 1890, 1894, and others in more recent years. These are dealt with later. Jardine mentions in this edition that Professor Bell already had in hand a new edition. This did not appear until 1877.

RE-ISSUE OF JESSE AND JARDINE'S EDITION IN 1854. Bohn's Illustrated Library. A re-issue was made in 1854 and in this an erratum is shown on page iv, referring to "page 18, bottom line, for *bark* read *whole.*" This erratum was overlooked in the 1870 and 1872 editions. It referred to a passage in one of Jesse's notes. Red cloth.

Second-hand: 3*s.*, 2*s.* 6*d.*, 7*s.* 6*d.*, 10*s.* 6*d.*

RE-ISSUE OF JESSE AND JARDINE'S EDITION IN 1861, 1867, 1870, 1872, AND 1881. Bohn's Illustrated Library. 5*s.*, or with coloured illustrations, 7*s.* 6*d.* Re-issues were made in the years mentioned. The publishers, however, now appear as Bell and Daldy, and the printer's name appears on the back of the title-page as W. Clowes and Sons., and also, in the case of the 1872 issue, on the last page of the book. Bound in green cloth.

Second-hand: 2*s.* (1870 ed.): 1*s.* 9*d.* (1872 ed.): 10*s.* 6*d.* (1854 ed.).

Re-issued by George Bell & Sons in 1878, 1888, 1890 and 1900.

# CHAPTER XI

## EDITIONS OF J. G. WOOD, BELL & DALDY, HARTING, BELL, & CHRISTOPHER DAVIES

REV. J. G. WOOD'S EDITION OF 1854. Issued without a date. Similar to that of 1860, but publishers are "London, George Routledge & Sons, New York: 416, Broome Street." 8vo. 3,000 issued. Price 3*s*. 6*d*.

It commences with a "Preface" and a "Biography". One impression of this issue was issued with the date, 1854.

REV. J. G. WOOD'S EDITION OF 1860.

Prior to the title-page, appear two illustrations, (1) "Queen Anne viewing the red deer in Wolmer Forest; (2) "Wolmer Pond," with "White's Selborne" at the top of the page, and the publishers' name and date at the foot.

Title-page. "The Natural History of Selborne. By "the late Rev. Gilbert White, M.A. With additional "notes, by the Rev. J. G. Wood, M.A., author of the "Illustrated Natural History, etc. Illustrated with en-"gravings on wood. London: Routledge, Warne, & "Routledge, Farringdon Street. New York: 56, "Walker Street, 1860."

The Preface is dated "Merton College, Dec. 20, 1853."

The volume commences with the poems, "Invitation to Selborne," "Selborne Hanger. A Winter Piece. To the Miss Bxxxxxs," "On the Rainbow." "A Harvest Scene," "On the dark still, dry, warm Weather," etc.

Page 9. "The Natural History of Selborne, in a series

of letters addressed to Thomas Pennant, Esq., and the Hon. Daines Barrington. Part I—To Thomas Pennant, Esq."

Letter I on page 11 is headed by an illustration of the beech. A picture of the elm appears on page 15, and there are many other small illustrations. The fossil oyster appears on page 19, although the original paragraph referring to the engraving of it is wanting. The Rev. J. G. Wood must have made his "copy" from an imperfect edition. See note on "Mytilus," page 113.

Part II commences on page 158, and contains the Letters to the Hon. Daines Barrington.

Between pages 236 and 237 is inserted a full-page woodcut of a gipsy girl under a rude tent.

"Observations on various parts of Nature, from Mr White's MSS. With remarks by Mr Markwick," appears on page 335.

Page 396. "Meteorological Observations."

Page 399. "Summary of the Weather."

Page 407. "A Comparative View of The Naturalist's Calendar," as kept at Selborne in Hampshire, and at Catsfield, in Sussex. Pp. 423-8, Index.

This is to some extent a bowdlerised edition, and was first issued in 1854.

Of Letter XVII to Pennant a paragraph and a half are omitted.

Letter XXXVI to Pennant. Two and a half lines are omitted at the end of the first paragraph.

Letter XXIV to Pennant. At the end, are added the words, "Disbelieving, I detest," a translation of "Incredulus odi."

Letter XIII to Barrington. Line 10 is omitted.

Letter XXI to Barrington. Fifteen lines are omitted.

Letter XXXII to Barrington omitted. Hence Letter XXXIII is called XXXII, and so on, all the subsequent numbering having been altered.

Letter XXXIII to Barrington. A few lines are altered.

A further edition of Wood's was issued in 1860 and 1875.

Routledges' informed me that they issued 2,000 copies of the editions of 1857, 1859, 1864, 1869, 1872, 1876, and 1879, at 3*s.* 6*d.* In 1893, 1,000 were issued at 3*s.* 6*d.*, and 2,000 at 2*s.*

16MO. EDITION OF 1862. Bell and Daldy's Pocket Library. Published at 3*s.* 16mo. "Antiquities" omitted.

Title-page. "The Natural History of Selborne, by "the late Gilbert White, A.M., Fellow of Oriel College "Oxford. With Miscellaneous Observations and Ex-"planatory Notes." [A woodcut of a Bell and an Anchor] "London: Bell and Daldy, 186 Fleet Street, and "Sampson Low, Son, and Co., 47 Ludgate Hill. 1862."

Preface; pp. v and vi.

Letters, numbered I to CX, pages 1 to 420.

Index, pp. 421 to 426.

Advertisement of the Pocket Volumes, 2 pages. At foot of p.426: "Chiswick Press:—Whittingham and Wilkins, Tooks Court, Chancery Lane."

In this edition, the two series of letters are printed chronologically, in one series only. The dates appear at the end of the letters. The "Miscellaneous Observations" have been interpolated by the editor (unknown) in various places, where they appear to be appropriate to the Letters. They are enclosed within brackets. Some of Mitford's notes are also quoted. There are many notes by "Ed."

I quote the following passage from the Preface:—

"Great as the interest of the letters themselves may be; and important as are the observations; these are not the only merits of the author. To him belongs the honour of having roused the observing faculties and directed the intellect of his countrymen to note and record natural phenomena coming under their notice. His humble and honourable career suggested a new path for research which was within the reach of thousands; to his example we probably owe the works of our Montagues, Selbys, Kirbys, Knapps, and Watertons. His example helped to substitute for the old "Book Naturalists," who had been the laureates of science from the days of Pliny, a race of observers who only recorded what they had seen for themselves."

"It must not be supposed that we mean any reproach to these authors; their merits are beyond praise, and the world is deeply indebted both to their philosophy and their facts. But it was a philosophy founded chiefly on analogy which, as the author remarks in these letters, 'is a very imperfect basis for Natural History.'"

We are informed that this edition was printed from the edition of 1822.

A re-issue was made in 1868 in two volumes. Second-hand: 1*s.* 6*d.*

HARTING'S EDITION OF 1875. Published by Bickers & Son. Printed by Whittingham and Wilkins.

For further particulars, see edition of 1876 (2nd

edition of Harting's), which contains in addition ten Letters to Marsham.

HARTING'S SECOND EDITION OF 1876.

Title-page. "The Natural History and Antiquities "of Selborne, in the County of Southampton. By the "Rev. Gilbert White, M.A. The Standard Edition by "E. T. Bennett. Thoroughly revised, with additional "notes, by James Edmund Harting, F.L.S., F.Z.S." An ornamental design, composed of honey-suckle. "Second Edition, with ten letters not included in any "previous edition of the work. Illustrated with en-"gravings by Thomas Bewick and others. London: "Bickers & Son, 1, Leicester Square, 1876."

Back of Title-page—"Chiswick Press:—C. Whittingham, Tooks Court, Chancery Lane."

Page iii, Publisher's Preface to the Second Edition, having reference to ten new letters from White to Robert Marsham which are inserted in this edition.

Pages v and vi, Contents. Page vi, an engraving of the Hermitage reduced from the larger plate in the edition (the second) of 1802.

Page vii, List of Illustrations, sixty in number.

Pages ix to xx, Preface by the Editor. On page xiii is an illustration, "Back view of the residence, at Selborne, of the Rev. Gilbert White." This is a reprint of that on page ix of Rennie's Edition of 1833.

The Preface is dated and signed on page xx, "Lincoln's Inn Fields, Sep., 1874."

Pages xxi and xxii contain Gilbert White's own preface, entitled "Advertisement to the First Edition."

Letter I is headed by an engraving of "Selborne from

Dorton." This is a reprint of that on the 13th page of Rennie's edition of 1833, but it is there entitled "The Village," a somewhat inappropriate title. Six lines from the "Invitation to Selborne" then follow.

Gilbert White's footnotes are initialled "G.W."

On page 7 is a woodcut of the Raven.

On page 9 is an illustration of the fossil *ostrea carinata*.

On page 13 is a reprint of an engraving entitled "Hollow Lane and Bridge, near Norton," which is called on page 13 in Rennie's edition "The Little Bridge."

On page 17 is another reprint from the same source entitled "Approach to the Village," but in Rennie's edition it appears as "Approach to the Village of Alton.

There are also many illustrations of birds and other animals, including one of the otter on page 97, and one of the peacock on page 110.

Letter I to the Hon. Daines Barrington, commences on page 136.

A "Sand Martins' colony at Oakhanger" appears on page 198. "Hawkley Slip" appears on page 263.

"More particulars" of the Tortoise are incorporated in Letter L on page 277.

On page 315 commence "Observations on various parts of Nature, from Mr. White's MSS., with remarks by Mr. Markwick and others."

Page 317. Observations on Quadrupeds.

Page 319. Observations on Birds.

Page 341. Observations on Insects and Vermes.

Page 355. Observations on Vegetables.

Page 357. An illustration of "The Grindstone Oak, in the Holt Forest."

Page 363. Meteorological Observations.

Page 368. Measure of Rain in inches and hundredths.

Page 369. Summary of the Weather.

On page 377, "A Naturalist's Calendar . . . by John Aikin, M.D., with remarks by Mr. Markwick and others." On page 379 is "Dr. Aikin's Advertisement," with a footnote by "Ed." (Harting).

On page 381, is "The Naturalist's Calendar, as kept at Selborne . . . . ." and "at Catsfield, near Battle in Sussex . . . . ."

Between pages 382 and 383 is a folding sheet, being a "Facsimile of a page from the Journal of the Rev. Gilbert White."

Page 383 contains "Preface to the Naturalist's Calendar," partly by [J.A.] the remainder being quoted from the edition of 1802, by J. White.

Page 385. "A Comparative View of White's and Markwick's Calendars." Ends on page 402.

Page 403. Title-page to "The Antiquities of Selborne, in the County of Southampton." Followed by the Arms of the Priory in a pointed oval, surrounded by the inscription, "Beate Marie D'Seleburne, Sigiil Convantus Eccle." A quotation from Virgil.

Letter I of the Antiquities commences on page 405.

The illustrations in this part of the book are few. On page 411 is a "South View of Selborne Church," the same as appears on page 456 of Rennie's edition.

"The Vicarage House" appears on page 419 (page 464 of Rennie's edition); "Temple in the Parish of Selborne" on page 438, (page 482 of Rennie); "The Plestor" on page 440 (page 6 of Rennie).

"The way leading to Gracious Street" appears on

page 512. The same illustration appears on page 3 of Rennie's edition, and is there entitled "The Hanger."

Page 515. Title-page of "Poems, selected from the MS. of the Rev. Gilbert White."

Page 525. "Appendix. Ten Letters from the Rev. Gilbert White, M.A., to Robert Marsham, F.R.S., 1790-1793." Pages 527-8, "Editor's Note," dated August, 1876.

Index, pages 561 and 562 (numbered 526) to 586 (numbered 532).

"Bickers & Son's List."

From the preface to this edition we find that Swan, Sonnenschein & Co., had acquired the copyright of the edition by Edward Turner Bennett (1837 ed.,) and that opportunity was taken to revise Bennett's notes, and to illustrate the work by some of Bewick's engravings. Many of the neat little engravings which illustrated both Rennie's edition of 1832, and afterwards Bennett's edition, are reprinted here, notably that of Gilbert White's residence at Selborne, as it stood when occupied by him. In this edition of Harting published in 1876 ten letters of White's are added, which were written to Mr. Marsham, of Stratton, Norfolk, a Fellow of the Royal Society. The first of these letters appears to have been a reply to a letter written to him by Marsham in 1790, which letter was discovered amongst White's MSS. after his death, and was published under Dr. Aikin's care in the volume of 1795 as part of "The Naturalist's Calendar." As Harting points out, on page 559, the last of this series of letters was written but eleven days before the writer's death. In it he speaks of having been annoyed during the spring with a bad nervous cough, and a wandering gout, which had "pulled me down very much." He was then in his 72nd year, and died on June 26th, 1793.

Second-hand: 3*s.*, 8*s.*, 3*s.* 6*d.*, 2*s.* 9*d.*, 9*s.* 6*d.*, 6*s.*

Bickers & Son published a third edition in 1880.

HARTING'S EDITIONS OF 1887, 1888 and 1889, Published by Swan, Sonnenschein.

Second-hand: 4*s.*

HARTING'S EDITION OF 1891. "Stereotyped Edition. Published by Swan, Sonnenschein, & Co., Paternoster Square."

Similar to that of 1875.

Second-hand: 1922, 6*s.*: 1932, 6*s.* 6*d.*,

There was a further issue with a printed slip attached to the title-page, stating that the book was now published by George Allen & Co., dated April, 1911.

THOMAS BELL'S EDITION. TWO VOLUMES: Demy 8vo. 1877.

In 1877 Professor Bell's edition was published. It is beautifully illustrated, and is dedicated to Lord Selborne. Some of Gilbert White's handwriting is reproduced in facsimile, and four new poems appear which are not met with elsewhere.

Bell's edition is in two Demy 8vo. volumes. The first contains the whole of White's own work, prefaced by a memoir, forty-two pages in length, of the author. For thirty-eight years Bell resided at Gilbert's own house at Selborne, and was well qualified to write concerning both author and village. The only notes to the letters which are quoted from a previous edition are those of the Rev. J. Mitford, in the edition of 1813.

The second volume contains a rich harvest of material which had been collected by the editor, throwing numerous sidelights on the character of Gilbert White. It includes correspondence with his brother, the Rev. J. White, of Blackburn, formerly of Gibraltar; correspondence between the Rev. J. White and Linnaeus; a suppositious letter from Timothy the Tortoise to Miss Mulso; correspondence with other gentlemen; a sermon; account book; garden calendar; geology by William Curtis; an appendix on Romano-British Antiquities by Lord Selborne, etc. The numbers of the pages in the two volumes are 507 (excluding the memoir, which is numbered separately) and 410 respectively.

*Volume I.* The title-page is prefaced by two leaves, on which are illustrations (1) "South view of the Church, with Yew and Vicarage;" (2) "Garden View of Gilbert White's House" (bay window and library added by Bell).

Title-page. "The Natural History and Antiquities of "Selborne, in the County of Southampton. By the late "Rev. Gilbert White, formerly Fellow of Oriel Col- "lege, Oxford. Edited by Thomas Bell, F.R.S., F.L.S., "F.G.S., &c., Professor of Zoology in King's College, "London. Volume I. Natural History, Antiquities, "Naturalist's Calendar, Observations on Various "Parts of Nature, and Poems. London: John Van "Voorst, 1, Paternoster Row, MDCCCLXXVII."

Back of title-page, crest and motto. "Printed by Taylor & Francis, Red Lion Court, Fleet Street."

Page v. Dedication to the Right Honourable Lord Selborne, F.R.S.

Page vii. List of illustrations in Volume I, eight in number. Page viii. "Corrigendum."

Pages ix to xv. Preface by "T.B."

Three aspects of *ostrea carinata* (fossil) as a tail-piece.

Inset is a reproduction of Gilbert White's handwriting (Latin and Greek quotations) with his signature.

Pages xvii to lix. "Memoir;" the Wych Elm as a tail-piece. Letters to Thomas Pennant occupy pages 1 to 112.

An illustration of the lovely walk down the Lythe (or Lith) appears opposite page 72, Dorton Cottage being in the centre of the picture.

Letters to Barrington occupy pages 113 to 274.

Bell makes a strange mistake on page 219, following Jardine in this respect: the remainder of the letter about the autumnal crocus *will* be found in the first edition, although, out of place, at the end of the "Antiquities."

An illustration of the Plestor appears facing page 310.

The "Antiquities" commence on page 275, on which appears the Seal of Selborne Priory.

At the end of the "Antiquities," on page 373, is placed "More particulars" of the tortoise.

Appendix: pages 375-403.

"A Comparative View of the Naturalist's Calendar," appears on pages 405-419.

"Observations on various parts of Nature" appears on pages 421-479.

The Rev. J. Mitford's notes (edition of 1813) are printed on pages 480-485.

"Summary of the Weather" occupies pages 487-495.

"Poems," 497-507. "The Invitation to Selborne;" "Selborne Hanger;" "On the Rainbow;" "A Harvest Scene;" "On the early and late blowing of the Vernal and Autumnal Crocus" (as well as on page 219, Letter XLI); "On the dark, still, dry, warm weather," etc.; "The Rhodian Children, as they go a'swallowing;" "On the late Armament, 1787;" "On a bad road, ill mended;" "A Fragment of a Satire." On the back of last page is the name of the printer, "Taylor & Francis."

*Volume II.* Illustration to half-title, "Sheepfold." Illustration facing title-page, "Front view of The Wakes—Gilbert White's House."

Title-page, same as to Volume I, except after the words "Volume II." Correspondence, Sermon, Account Book, Garden Kalendar, Animals and Plants, Geology, Roman-British Antiquities, etc.

Name of printer on back of title. List of illustrations to Volume II, six in number, on following page.

The volume opens with "Letters to his brother, the Rev. John White," which extend to page 64.

"Letter from John White to his brother Gilbert" appears on pages 65-66.

"Correspondence of the Rev. John White and Linnaeus" appears on pages 67-94.

"Correspondence with his family" appears on pages 95-185, the last letter, however, being a suppositious letter from "Timothy the Tortoise to Miss Hecky Mulso," dated "From the border under the fruit wall, Aug. 31, 1784."

An illustration of the west portion of the Village, on the road to Newton, faces page 138.

"Correspondence of the Rev. R. Churton and Gilbert White" appears on pages 186-230.

"Miscellaneous Letters" appear on pages 231-242.

"Correspondence of Robert Marsham, of Stratton Lawless, in the County of Norfolk, and Gilbert White" occurs on pages 243-303.

"On the Sense of Hearing in Fishes," pages 304-7.

A "Sermon" of Gilbert's is printed on pages 308-15.

"Gilbert White's Account Book" on pages 316-346.

"Gilbert White's Garden Kalendar," on pages 347-359.

"Description of Dufour's Fire Escape," on pages 360-1.

"List of the more noteworthy Animals and Plants observed in Selborne and the neighbourhood," pages 362-373.

"Geology of Selborne," by William Curtis, Esq., appears on pages 374-7.

"Appendix, on the Romano-British Antiquities of Selborne, by Lord Selborne," pages 378-394.

"Note on some recent discoveries in Selborne "Church, and their bearing on the History of the "Sudington Preceptory of the Knight's Templars." pages 395-8.

Illustrations (1) of Coffin-lid, opposite page 396; (2) of encaustic tiles from the Priory, opposite page 398.

Index, pages 399-410.

A lithographed copy of a page of White's Account Book is interpolated after the last printed page of the Index.

Second-hand: 18*s*., 10*s*. 6*d*., 12*s*. 6*d*., 11*s*. 6*d*., 31*s*. 6*d*.,

21*s.*, £2 10*s.* (1887): £2 2*s.* (1896): £2 4*s.* (1900): £1 5*s.* (1915).

A LARGE-PAPER EDITION was also published, with plates on India paper, at £3 3*s.*

Second-hand: £1 14*s.* (1887): £3 12*s.* (1889): £4 (extra bound, 1891): £1 15*s.* (1892): £1 18*s.* (1895): £1 16*s.* (1898): £1 5*s.* (1907): 16*s.* (1911): £1 (1930).

EDITION WITH NOTES BY G. CHRISTOPHER DAVIES (1879).

Frederick Warne & Co. (Chandos Classics). No date given. Published at 7*s.* 6*d.*

Illustration opposite title-page, entitled "View near Selborne;" locality not stated, but probably in Wolmer Forest.

Title-page on page iii. "The Chandos Classics. The "Natural History of Selborne, and the Naturalist's "Calendar. By the Rev. Gilbert White, A.M. A New "Edition, edited, with notes, by G. Christopher Davies, "Author of 'The "Swan" and Her Crew,' etc. etc. "Fully illustrated. London: Frederick Warne & Co., "Bedford Street, Strand."

Pages v, vi, and vii contain White's original Preface, or "Advertisement to Original Edition."

Davies' Introduction appears on pages ix to xvi. On page x is stated that White kept no personal diary. This is an error.

On the same page it is stated that the sketch of his life which appeared in the 1802 edition was by his brother John. This brother died in 1781; the sketch was written by his nephew, the son of Benjamin.

Page xvii. Contents.

Page xix. "The Natural History of Selborne." The first Letter to Thomas Pennant commences on the 21st page from the beginning, or page 3 of the Letters.

The Notes, which are short, are well placed at the end of each chapter.

Illustrations appear, of the Beech on page 4, of the Wych Elm on page 7. There are numerous representations of Birds; The Weasel, page 51; Metamorphosis of Newt on page 61; Viper and Ringed Snake on page 62; The Heron, page 74; Otter, page 94; The Tortoise, page 159; The Gipsies, page 200, etc., etc.

Although from the title-page it would be presumed that the "Antiquities" were not to be found in this edition, yet on page 291 we find the beginning of "The Antiquities of Selborne," extending to page 380. There are no illustrations to this part.

Page 381. "Observations on Various Parts of Nature. From Mr. White's MSS. With remarks by Mr. Markwick."

Page 382. "Advertisement to the 8vo Edition, published in 1802, edited by John White, the brother (*sic*) of the Author." This is dated "J. W., Fleet Street, 1802."

The "Observations" have a few illustrations, *e.g.*, "The Great Northern Diver (*Colymbus glacialis*)" on page 395; "May-flies in Sunset Dance" (a striking picture) page 415; "Snails and Slugs," page 424. "Meteorological Observations" commence on page 433.

The Poem, "On the dark, still, dry, warm Weather, occasionally happening in the Winter Months," appears on pages 436–7.

"Summary of the Weather" commences on page 437.

On page 445 commences the well-known "Comparative View of the Naturalist's Calendar," as kept at Selborne and at Catsfield, from 1768 to 1793.

Page 459. Poems selected from the MSS. of the Rev. Gilbert White.

Pages 461-2-3, "The Invitation to Selborne." Pages 464-5, "Selborne Hanger, A Winter Piece. To the Miss Batties." Page 465, "On the Rainbow." Page 466, "A Harvest Scene."

Page 467, Index.

Page 470, List of Illustrations.

In reviewing this edition in the *Zoologist*, the reviewer remarks . . . "We are unable to understand how any necessity can possibly have arisen for another edition of 'White's Selborne,' three different editions having appeared within the last three years . . . ."

A further edition appeared in 1890, published by W. W. Gibbings. Other undated issues were made, one being published in the "Golden Classics" by Myers & Co.

Second-hand: 3*s*. 9*d*., 1*s*. 6*d*., 6*d*.

## CHAPTER XII

## EDITIONS OF FRANK BUCKLAND

Frank Buckland's Edition (Macmillan & Co.) was published first in 1875, in one volume, medium 8vo. at 31*s*. 6*d*. This edition was not reprinted.

Second-hand: 10*s*., 10*s*. 6*d*., 22*s*. 6*d*., 14*s*.: 1889, £1 1*s*.: 1920, 14*s*.

Frank Buckland's Medium 4to Edition of 1876. Two volumes £5 5*s*. A chapter on Antiquities by Earl Selborne (*ob*. 1895). Also issued in one quarto volume.

With the exception of the photographs, most of the illustrations are similar to those in the 8vo edition of 1879.

*Volume I*. Title-page is faced by a View of Selborne from the Hanger, and runs thus:—"Natural History "and Antiquities of Selborne, by Gilbert White. With "Notes by Frank Buckland. A chapter on Antiquities "by Lord Selborne. And new Letters. With photo-"graphs and engravings from drawings by P. H. Dela-"motte. In two Volumes.—Vol. I. London: Macmillan "& Co. 1876."

Name of printer in a crest on back of title-page, "Richard Clay, London."

The Preface is signed by "Frank Buckland, 37, Albany Street, Regent's Park, Dec. 17, 1875."

"The Invitation"—a poem—commences the book, and is followed by "New Letters" (see notes on 8vo edition following). Opposite p.xxii is a photograph of a portion of White's letter to Thomas Barker, dealing with the rainfall of 1790.

Illustrations, 78 in number, exclusive of 15 full-page photographs of birds "by the carbon process."

Unfortunately for those who like the arrangement of the letters as they appeared in the original edition, they are arranged by Buckland to run according to their dates, and the two series of letters are not kept distinct. Consequently the last letter is numbered CIX; it concludes on page 292.

Opposite the conclusion of the last letter is a good map of the Environs of Selborne.

The Naturalist's Calendar follows, and completes Vol. I.

*Volume II.* Opposite the title-page of Volume II a coloured representation of a Romano-British Vase, of enamelled bronze, found at Blackmoor, from a drawing by P. H. Delamotte.

The contents of this volume are:

(1) Memoir of Gilbert White, pages 3—16.

(2) Notes and Observations, pages 17—150.

(3) Antiquities, pages 153—247.

(4) Appendix on the Romano-British Antiquities of Selborne, pages 251—266.

(5) Index to the two volumes, pages 269—283.

There are 46 engravings in this volume, exclusive of 8 photographs of birds "by the carbon process."

Second-hand catalogues: £2 2*s.*, £3 6*s.*, £2 17*s.* 6*d.* Auction prices: 1898, £1 14*s.*: 1894, £1 19*s.*: 1895, £3 3*s.*: 1911, £1 12*s.*, £1 10*s.*: 1909, 19*s.*: 1906, £1 11*s.*: 1912, £1 16*s.*, £1 19*s.*, £1: 1915, £1 6*s.*, 10*s.*: 1916, 10*s.*: 1920, £2 7*s.* 6*d.*: 1921, £2 2*s.*: 1924, 26*s.* ,30*s.*: 1925, 24*s.*, 20*s.*, 28*s.*

A LARGE-PAPER EDITION, 1876, was sold in 1887 for

£2 13*s*.: 1889, £2 2*s*.: 1890, £2 10*s*.: 1892, £2 2*s*.: 1893, £1 11*s*.: 1894, £2 7*s*.: 1895, 38*s*.: 1921, £2 2*s*.: 1922, £3: 1927, £2: 1932, 12*s*. 6*d*.

FRANK BUCKLAND'S 8VO EDITION OF 1877. One Volume. A chapter on Antiquities by Lord Selborne. Part of title-page in red, with a wood-cut of the blue-tit. Numerous illustrations with a coloured plate and map: pp. 591. Macmillan & Co.

FRANK BUCKLAND'S CROWN 8VO EDITION OF 1879. This edition was reprinted in 1880, 1883, 1887, 1891, 1894 and 1897. See 1887 edition below.

FRANK BUCKLAND'S CROWN 8VO EDITION, 1887. Price 6*s*. A chapter on Antiquities by Lord Selborne, and numerous illustrations by P. H. Delamotte. Also New Letters.

This edition (crown 8vo) was first printed in 1879.

Page 1 contains the words "Natural History and Antiquities of Selborne;" page 2, the publishers' initials set in a device consisting of a cross on a dark scalloped background, in the four arms of the cross being a butterfly, three stars, a bee, and three acorns, respectively.

On page 4 is a full-page engraving of "View of Selborne from the Hanger."

Page 5 is the title-page, particulars of the edition as above, with "London, Macmillan & Co., and New York, 1887," appearing at the foot.

Page 6 contains the printer's name, "Richard Clay and Sons, London and Bungay."

Pages 7, 8, 9, 10, 11, and 12 contain Frank Buckland's preface, dated "37, Albany Street, Regent's Park, December 17, 1880."

Page 13 commences "New Letters;" the first "The Invitation to Samuel Barker" begins with four lines in Latin, and then "See, Selborne spreads her boldest beauties round."

On page 15 begins a letter to "Dear Sam," dated "Selborne, Nov. 3, 1774;" on page 17, a letter to his sister, Mrs. Barker, dated "Dec. 25, 78;" and on page 18, a letter to "Dear Niece Anne." Another to his niece on page 19, dated "Feb. 5th, 1785," on the subject of the rain and frost; and on page 21, a further one to Thomas Barker, Esq., on the same subject, dated "Jan. 1st, 1791."

Page 23. Contents. Page 25. List of Illustrations, continued on pages 26, 27, 28 and 29.

On page 31 is the title of the first portion of the book, "The Natural History of Selborne."

The principal illustrations relating to Selborne are as follows:—Well-Head, page 3; The Plestor, page 6; Gilbert White's House from the front, page 9; Rocky Lane leading to Alton, page 11; Cottages in the Village, page 12; Old Coach Road, page 15; Path by Long Lythe, page 21; The Lythe, full page, page 37; Gilbert White's House from the garden, page 55; The Wishing Stone, page 65; and numerous illustrations of birds and eggs.

From page 263 to page 276 is "A Comparative View of the Naturalist's Calendar, as kept at Selborne, in

Hampshire, by the late Rev. Gilbert White, M.A., and at Catsfield, near Battle, in Sussex, by William Markwick, Esq., F.L.S., from the year 1768, to the year 1793."

"Memoir of Gilbert White" is found on pages 279 to 283 inclusive, while from page 285 to page 350 is found a series of "Notes and Observations" by Frank Buckland.

Principal illustrations:—Man-Traps, page 289; Rabbit with deformed Teeth, page 290; Rat's Tail, page 294; Mouse and Oyster, page 297; Eggs of the common Snake, page 305; Wren's Nest between two Stoats, page 321; Mummy Monkey found in a Tree, page 331; The Vegetable Caterpillar of New Zealand, page 348.

"The Antiquities of Selborne" commence on page 351, and end on page 441.

Principal Illustrations:—Gilbert White's House from the Garden, page 354; Selborne Church from the Alton road, page 358; the South Aisle, page 360; Original Benches in the South Aisle, page 361; View of the Chancel showing Gilbert White's monument, page 363; Facsimile of Death Register shewing an entry made by Gilbert White on June 10th, 1793, the following one being made by the Rev. C. Taylor, the Vicar, who registered on July 6th the burial of the Naturalist himself on July 1st, 1793, page 364; View inside South Porch, page 367; the Yew Tree in the Churchyard, page 368; Gravestone to Gilbert White, page 371; Sundial in Gilbert White's garden, page 441.

Appendix on Romano-British Antiquities of Selborne, page 442. Principal illustrations:—Map of Selborne Parish, page 444; large sepulchral vase, page 446;

water vessel, page 448; drinking cup, bronze celt and rings, page 449; spear heads, page 451; sword and sword handles, page 455; vase containing coins, page 459; coins, page 461.

Index, page 467 to page 480.

Macmillans' advertisements.

The illustrations are exceedingly beautiful and original. The engravings are by Delamotte. An important chapter on the Romano-British Antiquities of the neighbourhood is by the late Earl Selborne. This edition is the one for the multitude. Buckland added much that might appear to some persons to be extraneous and unnecessary matter, but his notes are imbued with so similar a spirit to that of White's that one can read on without a break, and feel that though one is reading the work of two authors yet they are of kindred minds.

## CHAPTER XIII

## EDITIONS OF HAWEIS, JEFFERIES, BURROUGHS, AND GRANT ALLEN

HAWEIS' THREEPENNY EDITION OF 1875. Selections only. Published by Routledge, with introduction by Rev. H. R. Haweis, M.A.

HAWEIS' SIXPENNY EDITION OF 1886. 12mo. Published by G. Routledge and Sons in their "World Library," with an introduction by Rev. H. R. Haweis, M.A. A few letters are omitted.

"The observations on various parts of Nature" appear on pages 141–160.

RICHARD JEFFERIES' EDITION OF 1887. Published by Walter Scott, etc., in "The Scott Library." 1*s*. 6*d*. Also in the Camelot series, 1887, in red cloth, cut edges, and dark blue cloth, uncut edges.

The preface by Richard Jefferies occupies six pages. The Antiquities are omitted. No annotations by the Editor.

The "Natural History" is found on pages 1–288, and the "Naturalist's Calendar" and the "Observations" on pages 289 to the end: 366 pages in all.

On page xi Jefferies refers to the little "Surrey" parish of Selborne. Two pages further on, in the first sentence of the first letter, White speaks of it as in the county of Hampshire. Jefferies could not have read the opening letter very carefully.

Reissue in 1896. 2*s*. 6*d*. crown 8vo.

MORLEY'S SHILLING EDITION OF 1887, (two volumes at *6d.* each, cloth; *3d.*, paper covers). Published by Cassell and Company, edited by Henry Morley, LL.D. The Introduction is by Professor Morley.

The edition consists of the Natural History Letters only: 192 pages in each volume.

It was reprinted in 1892 and 1897.

ROUTLEDGE'S EDITIONS OF 1879 AND 1887, with Jardine's Notes.

IN "THE EXCELSIOR SERIES." Pages, xix and 475.

ROUTLEDGE'S 8VO. EDITION OF 1880 WITH JARDINE'S NOTES. This edition contains both the Natural History and the Antiquities. It follows to a great extent the lines of Jardine's 1853 edition. The folding view of Selborne of the 1813 edition is reproduced on a small scale, and the title-page of the 1853 edition appears, with the reversed "N" in the word Selborne.

Jardine's "Introductory Observations" appear, and end on page xix. White's "Observations" begin on page 377.

John White's sketch of his uncle's life is printed on page 378. An interesting "Introduction to the Naturalist's Calendar" appears on page 441, apparently also by Jardine.

There are seventy-six illustrations.

PAPER-COVER SIXPENNY QUARTO EDITION, (1882?) In Routledges' Sixpenny Series, no. 28. Paper covers:

quarto. With 160 illustrations, from Wood's editions: 64 pages. Notes are quoted from Markwick, Mitford, Bennett, and Jardine. 30,000 copies were issued; well printed; fully illustrated: a remarkable edition.

In a letter to the writer the late Professor Alfred Newton wrote in 1897, "I have not seen one [edition] that I approve, except perhaps that which was issued at the price of *6d.*" (This remark, may, however, refer to Haweis' edition of 1886.)

It contains the "Observations," and the Poems:—"The Invitation to Selborne," "On The Rainbow," and "A Harvest Scene."

R. Clay and Sons and Taylor, Printers.

ROUTLEDGES' ISSUE OF 1890. In the "Camelot Series." Also in the same firm's "Popular Library" of 1890.

ROUTLEDGES' ISSUE OF 1891. In the "Camelot Classics." Jardine's notes. Numerous illustrations.

ROUTLEDGES' ISSUE OF 1891 AND 1894. In "Lubbock's Hundred Best Books." Jardine's notes. (See details later, in the similar edition which is without date.)

ROUTLEDGES' ISSUE OF 1895. Price, *3s. 6d.*

IN LUBBOCK'S HUNDRED BEST BOOKS. George Routledge and Son. No date in this issue: Jardine's notes, with his introductory observations concerning the Naturalist. White is referred to as the "almost obscure Vicar," but he was never Vicar. A paragraph concerning the absence of a portrait of the Naturalist is quoted from *Notes and Queries*, as a footnote. The reversed "N" appears on the title-page (see page 140). 475pp., including Antiquities, Poems, and Index. The "Introduction to the Naturalist's Calendar" is evidently by Jardine, and he refers therein to quotations made largely from his ornithological calendar prepared in 1849.

EDITION OF 1894, WITH ILLUSTRATIONS BY HARRISON WEIR. This edition was published by Routledges, in 1894, with six coloured plates and numerous other illustrations. It was re-issued in 1904.

BURROUGHS' EDITION OF 1895. In two volumes. 8vo. 10*s*. 6*d*. A beautiful edition, exquisitely illustrated with 47 and 33 illustrations. Printing and paper excellent, but the binding, in pale-green, coarse-looking cloth, scarcely in keeping with the rest of the book. Volume I has 208 pages, exclusive of Introduction; Volume II has 231 pages.

Volume I. "Natural History of Selborne," stamped in gold on front outside, over a representation of a bird.

"White's House, 'The Wakes,' from the Garden," forms a full-page frontispiece. It represents the house as it was in 1895. The right-hand half represents the century-old portion of the house.

Title-page, page iii; a handsome and elaborate scroll of conventional wild roses with a waxwing singing. Legend:—"Natural History of Selborne and Obser- "vations on Nature, by Gilbert White. With the text "and new letters of the Buckland edition. Introduction "by John Burroughs. Illustrations by Clifton Johnson. "London and New York: Macmillan & Co., 1895."

Page iv back of frontispiece, "Printed in America. Copyright, 1895, by D. Appleton & Co."

Pages v and vi, "List of Illustrations. Volume I."

Pages vii—xxiii, Introduction by John Burroughs dated April, 1895, and containing illustrations of "Interior of Selborne Church," page x. "Under the great yew in Selborne Churchyard," page xiii. "Faringdon Church," page xv. "Selborne Church from the short lythe," page xiv. "A Faringdon 'by-way,'" page xviii. "White's grave," page xxii. "A stile in the long lythe," page xxiii.

"Selborne from the Hanger" faces Letter I, which appears on page 1 of the book.

Notable illustrations are "Wolmer Forest," facing page 18, "Stonehenge," facing page 81, "Beeches in the 'long lythe,'" facing page 96. "A cottage in the lythe," facing page 98. "In a Selborne beechwood," facing page 183. "White's sun-dial at the bottom of his garden," facing page 202. "A rookery," facing page 207.

On page 204, "Bramber" is mis-spelt "Brambler."

On page 206, "Brightelmstone," as it appears in the original, is modernised to "Brighton."

Volume II. Similar frontispiece to Volume I, but faced by illustration of "Spring in a Selborne garden." Pages v and vi, "List of Illustrations."

On back of frontispiece, same as in Volume I, page iv.

Letters commence with Letter LVII to the Hon. Daines Barrington.

Notable illustrations are "Selborne cottages," page

9. "A look across the Plestor," facing page 37. "A pond on Selborne hill," facing page 68. "Plowing under the Hanger," page 127. "Burning an old hedge under the Hanger," facing page 141; this must have been near the northern end of the Hanger, where the Down slopes down by the hop-fields. "Selborne Down," facing page 171.

"New Letters" appear on pages 199—214.

"A comparative view of the Naturalist's Calendar" occupies pages 214—231.

In this edition the "Observations" are placed in brackets at the end of certain chapters to which they appear applicable.

The Letters are not numbered as in the original edition, but in chronological order.

BLACKIE'S SCHOOL AND HOME LIBRARY, 1895. Price 1*s*. 4*d*. pp. 252.

Title-page. "The Natural History of Selborne and "The Naturalist's Calendar By Gilbert White a New "Edition London Blackie and Son, Limited, 50 Old "Bailey. E.C. and Glasgow and Dublin."

Page 5. "Contents. Page 7, "Introduction", short but admirable: unsigned. Page 10, "The Invitation to Selborne," page 13, "The Natural History of Selborne" page 227, Observations on Nature"; page 239, "Comparative View of The Naturalist's Calendar as kept by Gilbert White, and William Markwick."

The Notes are wisely few. The whole is neatly and clearly printed.

In the same year the edition was also issued as one of the "Famous Books for Boys and girls," at 1*s*. 6*d*.

Some years later (I bought a copy in 1907) the same

edition was issued at 9*d*. net, in "Blackie's Library of Famous Books." An excellent feature of the edition was that White's numerous quotations from classic authors were translated for the readers, the translation appearing as footnotes. Some of the "Observations on Nature" are introduced at appropriate places in the text of the Letters. A footnote on page 27 says, "See Letter XLI to Mr. Barrington," respecting certain "curious plants." The list of plants is in this edition contained in Letter XL to Barrington. A letter has evidently been omitted, as the list certainly appeared originally in Letter XLI.

Ellwanger. Aldine Edition. 1898. In the World's Great Books. D. Appleton & Company, New York.

"With a Critical and Biographical Introduction by George H. Ellwanger."

One volume, Large 8vo. Pages, xx and 381. With two photogravures and five colour plates.

Contains the Natural History, the Antiquities, and Observations.

Grant Allen's Foolscap Quarto Edition of 1899 Price 21*s*., and $7.50. "Antiquities" omitted. A Photogravure portrait of Thomas Pennant appears as a frontispiece.

Title-page. "The Natural History of Selborne by "Gilbert White. Edited with Notes by Grant Allen. "Illustrated by Edmund H. New." Passages are quoted from Homer and Cicero. "John Lane The Bodley

Head London and New York 1900" (First published in monthly parts commencing in Oct. 1899).

Page vi. "Printed by Ballantyne, Hanson & Co. At the Ballantyne Press."

Page vii. "To A.E.L. in commemoration of August 13, 1898 this edition is dedicated by the publisher."

Page ix. "Notes." Acknowledgement by the Artist.

Page xi. "Contents."

Pages xiii to xxi. "List of Illustrations."

Pages xxii and xxiii. An old map of "The Village of Selborne."

Pages xxiv and xxv. An old "Map of Selborne and the neighbouring villages."

Page xxvi. "The Plestor."

Pages xxvii to xl. Introduction by Grant Allen.

Page 1. "The Natural History of Selborne."

Page 2. Illustration, "Selborne from the Hanger."

Pages 3 to 159. Letters to Pennant.

Pages 160 and 161. "North East view of Selborne from the Short Lythe in Gilbert White's time."

Page 162. White's Tombstone.

Pages 163 to 389. Letters to Barrington.

Facing page 163 is a portrait of the Hon. Daines Barrington.

Page 390. An illustration of a pair of candle snuffers.

Page 391. "White's Monument."

Page 392. Blank.

Page 393. "Observations" etc.

Page 394. Blank.

Pages 395 and 396. John White's advertisement of 1802.

Pages 397 to 454 Observations.

Pages 454 and 455. The Poem "On the dark, still, dry Weather" etc.

Pages 456 to 464. "Summary of the Weather."

Pages 465 to 485. "The Naturalist's Calendar."

Pages 487 to 494. Poems, including "On the dark, still, dry Weather" etc.

Pages 495 to 500. "Appendix," containing some notes by S. T. Coleridge.

Pages 510 to 516. Bibliography.

Pages 517 to 528. Index.

Four pages of advertisements.

Grant Allen omits in this edition the passage headed "More particulars respecting the old family tortoise, omitted in the Natural History." The omission was doubtless caused by the fact that in the original edition it appeared at the end of the "Antiquities," and this portion of the book was omitted altogether in this and many other editions.

The poem "On the dark, still, dry, warm Weather," was printed twice, viz., with the Poems on page 493, and after the Meteorological Observation, on pages 454-5.

The passage, "Of all the Propensities of Plants," etc., on page 429 of the original edition, is printed by Grant Allen, but he remarks, "I do not know whence Mitford derived it." Evidently the copy on which Allen founded his edition was Mitfords' edition of 1813. He would have met with the passage had he examined the original edition.

Some MS. notes by Samuel Taylor Coleridge are reprinted from his copy of the 1802 edition, which is in the British Museum. They are contained in a Appendix, and are entitled "Marginalia from Samuel Taylor Coleridge's copy of White's Selborne, here printed for the first time, with the assent of Mr. Ernest Hartley Coleridge."

Some of Grant Allen's remarks bear close resemblances to some of mine in my *Bibliography*, published in 1896, but mention of my work is omitted from Allen's Bibliography.

Second-hand: 14*s*., 10*s*. 6*d*., 15*s*.

This edition was also issued in 8vo. in 1902 and 1908.

Second-hand, 1908 edition. 5*s*. 6*d*.

There are differences in pagination, and the photogravure plate facing the title-page contains portraits of both Pennant and Barrington.

# CHAPTER XIV

## EDITIONS OF POLLARD, BOWDLER SHARPE, WEEKES, MIALL AND FOWLER, KEARTON, AND OTHERS

POLLARD'S EDITION OF 1900. In "Library of English Classics." Large 8vo. A further edition was issued in 1905.

Page i. "Library of English Classics," "Gilbert White," in red. Page ii. Initials of the publishers.

Page iii. Title-page. "The Natural History and An- "tiquities of Selborne [red] in the County of Southamp- "ton By Gilbert White." Latin quotations from Horace and "Scopoli Ann. Hist. Nat." "London, Macmil- "lan and Co. Limited" [publishers' name in red] "New York: The Macmillan Company 1900."

Pages v and vi. "Bibliograhical Note," signed by "A. W. Pollard." In this the editor states that "the present edition contains a faithful reprint" of the Natural History and Antiquities and the Naturalist's Calendar and the Observations, being as the editor thinks "the first [edition] in which they have been thus brought together without any addition or diminution."

Page vii. "Contents."

Pages ix and x. "Advertisement" from first edition.

Page 3. "The Natural History of Selborne."

Pages 3 to 255. Letters to Pennant and Barrington.

Pages 257 to 382. Antiquities.

Page 383. The Naturalist's Calendar.

Page 405. Observations on Various Parts of Nature.

Page 461. Index.

Page 476. "Glasgow; Printed at the University Press by Robert Maclehose and Co."

BOWDLER SHARPE'S EDITION OF 1900. In two volumes, large 8vo. Cash Price, £3 nett. in Great Britain and Ireland. Large-Paper edition limited to 160 copies (of which 150 were offered for sale), signed by editor and artist, two volumes, £7 net. This edition is unique by the inclusion of the "Garden Kalendar," printed here in its entirety for the first time, with an introduction to the same by Dean Hole. Bound in fine green canvas.

*Volume I.* Page iii "Gilbert White's Selborne Volume I." Page v. Title-page, "The Natural History and "Antiquities of Selborne and A Garden Calendar" [the foregoing printed in red] "By the Reverend Gilbert "White, M.A. Edited by R. Bowdler Sharpe, LL.D., "with an introduction to the Garden Calendar by The "Very Reverend S. Reynolds Hole, Dean of Roches- "ter, and numerous illustrations by J. G. Keulemans, "Herbert Railton, and Edmund J. Sullivan. In two "Volumes. Volume one" [Coat of Arms] "London: "Published by S. T. Freemantle in Piccadilly. Anno "Domimi MDCCCC." The title-page is faced by a portrait of Thomas Pennant.

Page vii. "Contents."

Pages ix to xii. "List of full-page illustrations," fifty in number.

Pages xiii to xv. List of smaller illustrations, thirty-seven in number.

Pages xvii to xxiii. Introduction.

Pages 1 to 185. The Letters to Pennant. Between pages 44 and 45, a photographic reproduction of Gilbert White's letter to Pennant, dated November 6th, 1767, is interpolated. This is an exact reproduction of the MS. letter of four foolscap pages, and is a good example of the whole series of original letters now in the British Museum. On the last page appears the address "To Thomas Pennant, Esq., at Donnington Flintshire, North Wales. A single sheet." There are also two old Post Office date-stamp impressions, and an endorsement in Pennant's writing.

Page 187. "A Garden Kalendar, edited by R. Bowdler Sharpe, LL.D., F.L.S. With an Introduction by The Very Rev. S. Reynolds Hole, D.D., Dean of Rochester."

Pages 189 to 421. The Garden Kalendar.

Pages 422 to 424. "Appendix I", consisting of a few belated notes to the Letters.

Pages 425 to 427. "Appendix II," consisting of a "Note on the Geology of the Selborne District," by C. W. Andrews, F.L.S. "End of Volume I."

*Volume II.* This appeared in 1901. Title-page as in Volume I, faced by a portrait of the Hon. Daines Barrington.

Page vii. "Contents."

Pages ix to xii. "List of full-page illustrations," forty nine in number.

Pages xiii to xv. "List of smaller illustrations," thirty two in number.

Pages 1 to 205. The Letters to Barrington. Between pages 158 and 159 is interpolated a photographic reproduction of the original manuscript of Letter LI to Barrington, dated April 22nd, 1780.

Page 207. Title to "The Antiquities of Selborne, in "the County of Southampton." Seal of the Priory, with the motto "...Juvat ire .... Desertosque videre locos.... Virgil."

Pages 209 to 314. "The Antiquities of Selborne."

Pages 315 to 316, "More Particulars respecting the old Family Tortoise omitted in the Natural History."

Pages 317 to 345. Appendices I to V.

Pages 345 to 364. "Bibliography."

Pages 365 to 443. "Index and Errata."

Second-hand; octavo, 28s., 42*s.*, 35*s.*, 26*s.*, 25*s.*, 14*s.*, £2 15*s.* (1928): £1 12*s.* (1932). Large-paper, £4 16*s.* (1907): £3 17*s.* 6*d.* (1910): £2 6*s.* (1911): £4 10*s.* (1914): £2 15*s.* (1921): £45 (special decoration and binding; signed by editor and artists; one of 160 copies): £3 (special binding): £3 3*s.* (1922) £5 5*s.*, £4 15*s.*, £6 10*s.*, £8, £4, £2 19*s.* (1926), £7 15*s.* (1928): £5 5*s.*, £8 (1929): £5 5*s.* (1930): £4 10*s.* (1931): £2 13*s.* (1933).

WEEKES' EDITION OF 1901. (Ed. I. Gollanz) 12mo. In the Temple Classics. Title-page faced by an excellent etching of the Hanger from Gracious Street.

Title-page. "The Natural History of Selborne. By "Gilbert White. MDCCCCI. Published by J. M. Dent "and Co., Aldine House London, W.C."

The Antiquities and Calendar are omitted.

Page iv. "List of Illustrations," fifteen in number, the first being the fossil oyster, and the remainder being various birds.

Page v. The Advertisement" White's original preface.

Page viii. Extract from Richard Jefferies.

Pages 1 to 336. The Letters.

Page 337. The Editorial Appendix consists for the most part of John White's short biography which ap-appeared first in 1802. Also contains some remarks as to the text used, viz., that of the first edition. Spelling has been modernised in a few instances.

Pages 341 to 371 contain the Notes, many of which are ,as it seems to me, superfluous.

Page 381 (back of). "Edinburgh, Colston and Co., Limited, Printers."

MIALL AND FOWLER'S EDITION, 1901. 8vo.

Title-page. "The Natural History and Antiquities of "Selborne by Gilbert White, Edited with an Introduct-"ion and Notes by L. C. Miall, F.R.S. and W. Warde "Fowler, M.A. Methuen & Co, 36 Essex Street W.C. "London 1901." The words "Natural History" and "Selborne" are printed in red.

The original edition of 1789 is reproduced, except the Appendix of Latin charters and deeds. The Naturalist's Calendar is not reprinted, as "its scientific value is small."

"Introduction," pages ix to xxxv, contains much useful and detailed information. The naturalist was not born at "The Wakes," as stated, but at the old Vicarage.

Reduced facsimile of original title-page, xxxvi. Advertisement to original edition, xxxix.

"The Natural History of Selborne," page 1. "The Antiquities," page 237. "Observations," page 319. Index page 357. Printer, The Aberdeen University Press, Ltd.

The editors' notes are useful, and to the point. On pages 165 to 167, the editors have made lengthy notes on the so-called "dew-ponds," a name by which, it may be mentioned, White nowhere calls them.

Second-hand: *2s. 3s. 6d.*

RICHARD KEARTON'S EDITION OF NOV. 1902. Price 6*s*. nett. Bound in green Canvas. Published by Cassell and Co. "White's Selborne" on back of cover. This edition is noticeable for its excellent illustrations of bird life. It contains only the "Natural History." The annotations are few and are unimportant. Facing the title-page is a beautiful carbon photograph of part of the village from a gap in the Hanger. The picture shows the modern portion of White's house, the Church, the Vicarage, and the line of trees which marks the Lith, and the stream on its way to Priory Farm.

Title-page. "The Natural History of Selborne, By "Gilbert White. With Notes by Richard Kearton, "F.Z.S., and Illustrations from Photographs taken "direct from Nature by Cherry and Richard Kearton. "Cassell & Company, Linited, London, Paris, New York, and Melbourne. MCMII. All rights reserved."

Page v. List of Illustrations. These are one hundred and twenty-four in number, and are excellently reproduced from photographs.

Pages vii to xiv. Introduction by the Editor.

Pages xv and xvi. White's original "Advertisement.'

Pages 1 to 284. The Natural History Letters.

Pages 285 to 294. Index.

8 pages of advertisements.

Second-hand: 2*s*. 6*d*., 3*s*., 5*s*. 6*d*.

Three reprints of this edition subsequently appeared, namely, in January and September, 1903, and in 1911. The last mentioned contains as a frontispiece a coloured full-page illustration of the "Kingfisher in Winter."

In "The World's Classics" 1902 Post 8vo. Volume No. XXII in the series. Cloth, 1*s*., Leather, 2*s*. net.

"The Works of Gilbert White. The Natural History of Selborne."

A page of advertisements of The World's Classics,

Title-page. "The Natural History of Selborne. By "the late Rev. Gilbert White, A.M. Fellow of Oriel "College, Oxford. (Publishers device "G.R.") London, "Grant Richards, Leicester Square, W.C. 1902."

On the next page we are informed, "The Natural "History of Selborne, was first published in the year "1789. In 'The World's Classics' it was first published in "1902." "Printed by R. & R. Clark, Limited, Edin-"burgh."

Page 1. "Poems selected from the manuscript of the Rev. Gilbert White."

Page 3. "Illustrations to Selborne." Page 5. "Selborne Hanger." Page 6. "On the Rainbow." Page 7, "A Harvest Scene." Page 8. "On the dark, still, dry Weather."

Pages 13 to 263. The Natural History of Selborne.

There is no Index. The Antiquities and the Observations are omitted. This little edition was issued by the publishers in a brilliant crimson paper guard-cover.

The "World's Classics" were later taken over by the Oxford University Press (Froude), and after the war the price was doubled.

The Union Library. Crown 8vo., cloth gilt: 2*s*. 6*d* : Walter Scott, Ltd.

Library of Standard Works. Leather, gilt edges: 4*s.*: Sunday School Union.

Morse's Edition of 1904. Published by Ginn & Co. 2*s.* 6*d.* 251 pages. Edited by E. S. Morse.

Windle's Edition of 1906. In "Everyman's Library" published by J. M. Dent & Co., Kewford Street, London, W.C. Cloth 1*s.*, net, Leather, 2*s.* net. Introduction by Professor B. C. A. Windle. Reprinted in 1909 and 1912.

Contains the Natural History letters only, with no annotations but White's. 255 pages. One illustration, on page 10, of *Mytilus, Crista Galli.*

Naturalist's Edition of 1906. Edited by Charles Mosley. 259 pages and Index. Crown 8vo. 6*s.* net. Elliot Stock.

The arrangement in this edition is new, in that the letters have been cut up and re-arranged under subjects. The charm of the original work is lost, but the edition served a useful purpose. No less than 134 pages are devoted to birds; meteorology claims 30 pages; mammals, 21 pages; and botany, 10 pages.

The folding North-East View of Selborne from the Short Lythe. which appeared in the first edition is reproduced on reduced scale.

Second-hand: 2*s.* 6*d.*, 3*s.*, 1*s.* 6*d.*, 3*s.* 6*d.*, 5*s.* 6*d.*

The World's Library of Famous Books; Ward, Lock & Co., undated.

Another edition in "The World's Library" was issued with the date 1912, in green binding: 12mo.

Miall and Fowler, in Methuen's Standard Library, 1906. Paper *6d.*, cloth *1s.*, net, edited by Sidney Lee.

Opposite to title-page is a rude and quaint symbol with the name of the Library, and the words "Arbor Scientiae."

Title-page contains two quotations from Horace and Scopoli.

Pages iii to xxxvii contain an Introduction, of which Sections i, ii and iii are by L. C. Miall, and Section iv by W. Warde Fowler. On pages xxxix and xl is the "Advertisement to the First Edition."

The Natural History follows on pages 1 to 186 and 189 to 257. Appendix, pages 259 to 279. Index, pages 281 to 286.

This edition does not contain the Observations, the Naturalist's Calendar or Poems.

Cassell's People's Library, 1908. No. 55, bound in green cloth.

An "Editor's Note" on page 4 faces the title-page.

Publisher's name on title-page, as "Cassell and Co., Ltd, London, Paris, and New York, MCMVIII."

"Advertisement "on pages 7 and 8. This is Gilbert White's original preface.

Illustration of*Mytilus* on page 15.

Natural History on pages 1 to 289.

Index, pages 291 to 299

There are a few notes.

No price is stated.

EDITION WITH INTRODUCTION BY H. G. GROSER, 1908. Andrew Melrose, Ltd, London. One volume, 8vo. pp. 167. With Explanatory Notes.

EDITION BY H. B. BROWNE, M.A., in "Arnold's English Texts." T. Arnold, London. One volume, 18mo. pp. 96; *6d.* N.D. In limp red cover. Thirty-one letters only are selected and printed in this edition.

SELECTIONS FROM WHITE'S NATURAL HISTORY OF SELBORNE, 1909. "English Literature for Secondary Schools." Notes by F. A. Bruton. General Editor, J. H. Fowler, M.A., Macmillan and Co., Ltd.

A picture of the village from the top of the Hanger appears opposite to the title-page.

Title; "Selections," etc., "edited with introductory notes and glossary by F. A. Bruton, M.A., Asst.Master at Manchester Grammar School, with Forty illustrations. Macmillan and Co., Ltd, St Martin's Street, London, 1909.

Introduction, pages xiii to xix.

Letters, selected, 1 to 126.

Notes, 127 to 131.

Glossary, 132, 133.

Questions, 134, 135.

Subjects for essays, 135, 136.

Short Bibliography, 136.

Bound in green limp cloth: price one shilling.

Although the extracts are arranged in the form of letters, thirty in number, the numbering of the Letters bears no resemblance to the numbering of the original Letters. It is a pity that reference numbers to these were not inserted.

The School Selborne, 1910. This edition consists of selections from The Natural History of Selborne, arranged and edited by F. H. Shoosmith, B.Sc. (Lond.) Charles and Dible, London, Glasgow, and Dublin.

A view of "Selborne from the Hanger" faces the title-page.

Preface, pages v and vi. Contents, vii and viii. "The School Selborne." pages 9 to 95. A few good illustrations. Bound in green limp covers: *6d.*, net.

Collins' Edition of 1911. Macmillan & Co., Ltd. With 24 Coloured Illustrations by George Edward Collins. Crown 4to. *10s. 6d.*, net.

Contains the Natural History, pages 3 to 255; Antiquities, page 257; Naturalist's Calendar, 383; Observations, 405; Summary of the Weather, 451; Index, 461.

Title-page (faced by a coloured illustration of "Selborne from the Hanger") "The Natural History and "Antiquities of Selborne in the County of Southampton, By Gilbert White: with illustrations in colour by "George Edward Collins, R.B.A. Macmillan & Co., "Ltd, St Martin's Street, London, 1911."

White's "Advertisement," pages ix and x. "More particulars of the Family Tortoise." follows after Letter XXVI of the Antiquities. The poem, "Say, what impels," etc. appears at the end of Letter XLI to Barrington. The peculiarities of a "hollow, rocky lane" are well shown in the illustration facing page 11.

A reviewer in *Nature* said of this edition that it "is free from editorial interference, and the result is a beautiful edition." It contains no editorial notes.

The volume is bound in green cloth, with a picture of Selborne Street stamped in gold on the cover.

EDWARD STEP'S EDITION, 1921. In The King's Treasures (general editor, Sir A. T. Quiller-Couch). Abridged and edited by Edward Step, F.L.S. One illustration, "Selborne, The Hanger." pp. 256. J. M. Dent and Sons, Ltd.

KEARTON'S EDITION OF 1924. Quarto. Arrowsmith, 21*s*. Contains the Natural History only. Re-issued in 1931.

Title-page, "The Natural History of Selborne by "Gilbert White, with notes by Richard Kearton, F.Z.S., "and 85 photographs, including many specially taken "at Selborne by Cherry Kearton and Richard Kearton. (Here a woodcut illustrating a man hammering an arrow on an anvil). "Arrowsmith, London, W.C.1."

Title-page faced by picture of a "young Carrion Crow."

Page 4. "This edition was first published in 1924." "Printed by J. W. Arrowsmith Ltd, 11 Quay Street, Bristol."

Pages 5, 6, 7. List of "Illustrations."

Pages 9, 10. "Twenty years after," dated "R. Kear-

"ton, The Queens Hotel, Selborne, Hampshire, Eng-"land, May, 1924."

Pages 11 to 14. "Introduction," dated "Richard Kearton, Caterham Valley, November, 1902." The statement made that Gilbert White was born at "The Wakes" is incorrect.

Opposite pages 14 and 15 are illustrations of "Massed Foliage" and "Dorton Cottage, Selborne."

Pages 15, 16 contain White's original "Advertisement." Illustrations opposite pages 16 and 17, "The Village consists of a single street," and "Beech Trees on the Hanger, Selborne."

Pages 17 to 105. Letters to Pennant.

Pages 106 to 243. Letters to Barrington.

Pages 345 to 248. Index.

Second-hand: 12*s*. 1926, 10*s*.

"White's Selborne for Boys and Girls." 1927 and 1928. Edited by Marcus Woodward. With Bewick's pictures and eight coloured plates. Published by Basil Blackwell, Broad Street, Oxford.

The editor remarks that "through generations it has been a favourite school prize-book, though totally unsuitable in its original form."

"The Natural History of Selborne." With Introductory Chapter, Edited by E. M. Nicholson: with woodcuts by Eric Fitch Daglish and map of surrounding parishes. Pp. 416. 4to. Thornton Butterworth. 1929. £2 2*s*.

Auction Prices: 1930, £1 10*s*., £1 1*s*.

# ADDITIONAL BIBLIOGRAPHY

## HESPERIDES EDITION 1937

The | Natural History | of | Selborne | in the County of Southampton | by | Gilbert White | [*Sketch of The Wakes by E. H. New*] | London | Humphrey Milford | 1937 |

8°. Pp. viii. 300.

The Hesperides series. 500 copies printed at the University Press, Oxford. Typography designed by Bruce Rogers.

## NONESUCH PRESS EDITION 1938

The | writings | of | Gilbert White | of | Selborne | Selected and edited | with an introduction by | H. J. Massingham | Volume I [Volume II] London 1938 | The Nonesuch Press |

8°. Vol. I, pp. xxxii. 312 and map. Vol. II, pp. viii. 356, and folding map.

Designed by Francis Meynell and illustrated with wood-engravings by Eric Ravilious. Published March 1938. Price £3 10*s*. Edition limited to 850 sets.

## PENGUIN BOOKS EDITION 1941

Issued in June 1941, edited by James Fisher. Illustrations by Clare Leighton. No. 296 in series.

## JAMES FISHER'S EDITION 1947

The | Natural History | of Selborne | by | Gilbert White | [*rule*] | Edited and with an introduction by | James Fisher | Illustrated with wood engravings by | Claire Oldham | [*rule*] | London [ The Cresset Press | MCMXLVII |

8°. Pp. xxiv. 296.

The editor states in a note that this edition is based on one he edited for Penguin Books, but that the notes have been revised and increased and that he has added an introductory essay. New impression 1966, price 18*s*.

## EVERYMAN'S LIBRARY EDITION 1949

In 1949 the edition in this series issued first in 1906, edited by B. C. A. Windle, was replaced by a new edition edited, with an introduction and notes, by R. M. Lockley. This was reprinted 1950 and 1966 (price 12*s* 6*d*).

## LUTTERWORTH PRESS EDITION 1951

The Natural History | of | Selborne | by | Gilbert White | With drawings by | John Nash | Lutterworth Press | London | 1951 |

8°. Pp. 308. Foreword by John Lewis.

## FOLIO SOCIETY'S EDITION 1962

The | Natural History | of Selborne | by | Gilbert White | Edited with an Introduction | and Notes by W. S[idney]. Scott. | Drawings by John Piper | London | The Folio Society | MCMLXII |

8°. Pp. xii. 216. Cover design by John Piper. Issued in cardboard slip case.

## SCOTT'S EDITION 1950 (*The Antiquities only*)

The Antiquities | of | Selborne | in the County of | Southampton | [*double rule*] By Gilbert White | [*double rule*] Edited by W. Sidney Scott | with notes and appendices | illustrated with wash drawings | of Selborne made in the year 1776 by | Samuel Hieronymus Grimm | [*design*] | London | The Falcon Press |

8°. Pp. 268. Price 15*s*.

This is the first time *The Antiquities* have been issued separately. Many editions of *The Natural History* omit them on the grounds that they are of little interest (a view shared by White's friend, John Mulso), but there is much of value in them, particularly the curious account of the Bishop of Winchester's visitation to the Priory in 1387.

DANISH EDITION

GILBERT WHITE | Selborne | ET SOGNS NATUR OG FUGLELIV | (cut of a bird) | ILLUSTRERET AF MADS STAGE | (*rule*) | H. HAGERUP, KØBENHAVN | 1951

(On verso of T.P.) Oversat efter Gilbert White | "The Natural History of Selborne" | af | NIELS HAISLUND.

Pp. XI, 379.

SWEDISH EDITION

GILBERT WHITE | Naturen i Selborne | NATUR OCH KULTUR STOCKHOLM | (engraved view of Selborne)

(On verso of T.P.) THE NATURAL HISTORY OF SELBORNE | ÄR ÖVERSATT OCH KOMMENTERAD AV AXEL LJUNGBERG | BOKFÖRLAGET NATUR OCH KULTUR MCMLXIII | PRINTED IN SWEDEN, VICTOR PETTERSONS BOKINDUSTRI AB | STOCKHOLM MCMLXIII

(On verso of half-title) Med illustrationer av | GUNNAR BRUSEWITZ | Inledning av Knut Hagberg

Pp. XVII, 286.

FRENCH EDITION

(Vol I)

GILBERT WHITE | THE NATURAL HISTORY | OF | SELBORNE | LETTERS SELECTED AND EDITED | BY | G. MIALLON | Professeur agrégé au Lycée Henri-IV | With 12 illustrations by E. H. New and 2 maps. | DIDIER | PARIS | 1949

(Vol II)

GILBERT WHITE | THE NATURAL HISTORY | OF | SELBORNE | LETTERS SELECTED AND EDITED | BY | G. MIALLON | Professeur agrégé au Lycée Henri-IV | With 2 maps, and 12 Illustrations by E. H. New. | INTRODUCTION ,NOTES AND EXERCISES | DIDIER | PARIS | 1949
Vol. I: pp. 109. Vol. II: pp. 67.

THREE JAPANESE EDITIONS

(approximate transliterations only are given)

1. THE NATURAL HISTORY OF SELBORNE.
   Translated by BUNSHO JAGAKU.
   Foreword by Edmund Blunden.
   TOKYO: IWANAMI SHOTEN: 1949.
   Vol. I: pp. 199. Vol. II: pp. 296, 79.

2. THE NATURAL HISTORY OF SELBOURNE.
   Translated by YOSHIO YAMANOUCHI.
   NARA: YOTOKUSHA: 1949.
   Vol. I: pp. 177, with folding map of village, names in Japanese.
   Vol. II: pp. 271, with folding map of district, names in Japanese.

3. THE NATURAL HISTORY OF SELBOURNE.
Translated by TAIZO NISHITANI.
Published privately by NAO HAMAGUCHI in memory of his brother, the translator.
Not for sale.
MUKŌMACHI, near KYOTO, JAPAN, 1958.
Pp. XIII, 547.

*The additional bibliography is included with acknowledgement and thanks to THE BOOK COLLECTOR*

# INDEX


Aikins' Edition 100
American Editions 120
"Antiquities," Letters of the 24
Argyll, The late Duke of 82
Arnold, Sir Edwin 82
Auction prices of First Edition 97

Barrington, Daines, Correspondent 24
Barrington, Daines, Inventor of Calendars 19
*Belief, The Burden of* 82
Bell and Daldy's "Pocket Library" 146
Bell, Prof. Thomas, resides at "The Wakes" 60
Bell, Prof. Thomas, Vault in Churchyard 41
Bell's Edition 152
Bells at Farnham 20
Bennett's Edition 132
Blyth's Editions 129-132
Bohn's "Illustrated Library" 137
Book, The—
- letters copied for printing 31
- the copies auctioned 31
- first edition 90
- original manuscripts 29, 67
- publication first contemplated 25

Bowdler Sharpe's Edition 177
Brown's Editions 121-128
Browne's Edition 185
Bruton's Selections 185
Bryce, Professor James 82
Buckland's Editions 160-165
Buckland, Frank, Visit to Selborne 15
Burrough's Edition 169
Butcher's shop, Trees in front of 43

Calendar, Change in the 4
Carlyle, Thomas, Appreciation by 88, 89
Chapone, Mrs. 11
Church, Age of the 41
Church, Triptych in the 42
Churton, the Rev. R. 20
Coleridge, Samuel Taylor, Notes by 174
Collins' Edition 186
Common, The Selborne 53
"Constable's Miscellany" 112
Contemporaneous events 4
Curacy at Faringdon 2

Darwin, Charles 86
Davies', Christopher, Editions 157
Defoe, his account of Great Storm, 1703 37
"Delves House," Ringmer 72
Dew-ponds 180
Dorton 44
Dover's, Lady, Editions 119

Ellwanger's Edition 172
"Everyman's Library" 183

Faringdon, Curacy at 2
Farnham, Church Bells of 20
First Edition 90
*Flora Anglica*, White's copy of 46
Fossils of Selborne 56
Froude's Edition 182

*Gentleman's Magazine, The* 3, 31, 71
German Edition 98

*Gleanings in Natural History* 17, 86
Grant Allen's Edition 172
Gravestone, The 40
Groser's Edition 185

Hales, Dr. Stephen 5
*Hampshire Days*, Hudson's 46
Hanger Hill 46
Harting's Editions 147, 148
Haweis' Editions 166
Hermitage, The 48
Hollow Lanes 34
House Crickets 142
Houston, Matilda 87, 88
Hudson, W. H., at Selborne 46

Jardine's Editions 114, 115, 167
Jefferies' Editions 166
Jefferies, Richard 85
Jenyns' Edition 135
Jesse, Edward 86
Jesse and Jardine's Editions 137, 143

Kearton's Editions 181, 187

Letters, The original 70
Long Lithe 36
Lowell, James Russell 80, 81, 82
Lubbock's "Hundred Best Books" 168

Memorial Seat 47
Meyer's Edition (Berlin) 98

Miall and Fowler's Edition 180, 184
Mitford's Editions 103, 107
Mitford, Miss 85
Morley's Edition 167
Morse's Edition 183
Mosley's Edition 183
Mulso, John; Eulogy 30
Mulso, John, visits Gilbert White 23
Mulso, Miss Hester 11
Mulso, Miss Heeky 78, 155
*My Study Windows* 80

"National Illustrated Library" 139
*Naturalist's Calendar, A* 98
Naturalist, The 1
Nicholson's Edition 188

Oakhanger, Lane to 49
Oriel College 1, 2
*Our Village* 85

Pagination of First Edition, Errors in 91
Parents, Gilbert White's 1
Pennant, Thomas, Correspondent 24
Plarr, Victor, Poem by 83, 84
Playstow, The 37
"Pocket Library," Bell and Daldy's 146
Pollard's Edition 176
Priory, The 40
Priory Farm 44
Priory Tiles 44, 45

Rennie's Edition 116

| | |
|---|---|
| Ringmer | 24, 72, 78 |
| Rivulets, The two | 35 |
| | |
| Sarsen-stones | 47, 63 |
| "School Selborne, The" (Shoosmith) | 186 |
| Selborne Common | 53 |
| Selborne, Geology of | 54, 55, 56 |
| Selborne Village | 34 |
| Selborne, visited by Frank Buckland | 15 |
| Selborne, Water Supply of | 51, 52 |
| Sermons, Auction Prices of | 69 |
| S.P.C.K. Editions | 119, 120 |
| *Speaker, The* | 83 |
| Storm of 1703 | 37 |
| Step's Edition | 187 |
| Sundial | 62 |
| | |
| "Tenpenny Nails" in walls | 39 |
| Tiles of Priory | 44, 45 |
| Tortoise, the Family | 72-79, 155 |
| Trees in front of Butcher's Shop | 43 |
| Triptych in the Church | 42 |
| | |
| Vault of Professor Bell | 41 |
| | |
| "Wakes, The" | 57 |
| alterations in | 58 |
| Prof. Bell resides at | 60 |
| brick path in garden | 62 |
| sale of | 61 |
| Weekes' Edition | 179 |
| Weir, Harrison, Illustrations by | 169 |
| Well-Head, The | 35, 51 |

White, Benjamin and John: book catalogues 96
White, Gilbert
Birth 1
Education 1
Proctorship 2
Deacon's Orders 2
resides at "The Wakes" 2
Curacy at Faringdon 2
his Brothers 3
his Friends 8
his gentle manner 8
his visits to London 10
his Costume 12
as a Geologist 12, 54
a "small man" 12
his Reputation 13
his printed Sermon 14
Portraits 15, 16, 169
his Diary 17
his difficulty in writing 27
makes an Index 28
Swedish Appreciation 33
lines addressed to himself on becoming Author 33
as a Botanist 46
Relics of 63, 67
suffers from a Cough and Gout 152
suffers from Small-pox 2
Windle's Edition 183
Wishing Stone, The 47
Wood's Editions 144
Woodward's Edition 188
"World's Classics," The 182

*Yale Review* 88
Yew Tree in Churchyard 38

Zigzag, The 46-48